THE BIOTECH PRIMER TWO

Biotech Primer Inc.

THE BIOTECH PRIMER TWO

Next Generation Therapies Explained

Vaccines, Cell, Gene, Antibody, and RNA Therapies
for the Non-Scientist

THE BIOTECH PRIMER TWO

Next Generation Therapies Explained

Published by:
Biotech Primer Inc.
7006 Copeligh Rd.
Baltimore, MD 21286

BiotechPrimer.com
training@biotechprimer.com

Published by Biotech Primer Inc., Baltimore, Maryland, USA

Biotech Primer Inc. also publishes its books in a variety of electronic formats. Content that appears in print is also available in electronic books.

ISBN 978-1-5136-5505-5 (softcover)
ISBN 978-1-5136-5503-1 (eBook)

ABOUT BIOTECH PRIMER

Biotech Primer Inc. delivers customized training to help professionals understand the science, business, and regulatory processes essential to the biopharmaceutical, medical device, and molecular diagnostic sectors. Our industry experts develop and deliver customized courses, so you are assured our content is always up-to-date and relevant to your training needs.

When we opened Biotech Primer's doors in 2001, our sole focus was entry-level science education for the non-scientist, which helped clients bridge in-house communication gaps and develop more cohesive, well-trained teams for product development. Since then, we expanded to offer advanced training for all levels of experienced professionals, including scientists—an estimated 100,000 people trained worldwide to date. We help prepare companies to make strategic business decisions, navigate important regulatory hurdles, and move healthcare products from the bench to the bedside. To accomplish these goals, we offer a diverse range of learning, ensuring participants retain and put into practice what they learn.

WE OFFER:

- **Customized Training:** Organizations can choose from our signature courses—featuring core material from our extensive library of life science topics—or let us fully customize a course with completely original content specific to your organizations' needs.
- **Open Enrollment Signature Courses:** Register to take our prescheduled, one-day, or two-day signature classes offered throughout the year live.
- **On-Demand Micro Classes:** Learn anywhere, anytime, at your own pace from our library of 60-minute on-demand classes. Designed for individuals, bulk purchase discounts available for organizations.
- **The Primer:** Hundreds of one-pagers explaining the science behind today's headlines. Head over to Learn on our website and take advantage of our original content.

biotechprimer.com

PUBLISHER'S ACKNOWLEDGEMENTS

We are proud of this book and welcome your feedback at training@BiotechPrimer.com.

For more information on the basic science that drives the biopharma industry read The Primer at BiotechPrimer.com/blog and follow us on twitter @BiotechPrimer.
For a list of live and on-demand online courses visit us at BiotechPrimer.com

Author: Emily Burke,
Ph.D. Editor: Sarah Van Tiem

Illustrations: Michelle Leveille
Cover & Book Layout: Tara Price

USER'S GUIDE

Throughout the book, you will see a variety of callouts, each designed with a specific purpose in mind. Here is a guide to the intent of each:

INDUSTRY NOTE: Explains how the biopharma industry uses science to advance healthcare.

BIOPHARMA INNOVATION: Examines a specific product, explains how it works, and relates it to the science described in the main text.

COCKTAIL FODDER: A Biotech Primer classic, pull them out the next time you are stumped for conversation at your company holiday party!

GOING FURTHER: Takes a more in-depth look at a topic related to the main text.

TRICKY TERMS: Highlights the meaning and usage of especially important industry terms or terms that tend to be confusing to non-scientists.

GLOSSARY: All **bolded and underlined** words in the text are found in the glossary at the back of the book.

Happy reading!

CONTENT

CHAPTER 3: THERAPIES: HOW BIOPHARMA FIGHTS DISEASE 35

CHAPTER 4: IMMUNOTHERAPIES: HARNESSING THE POWER OF THE IMMUNE SYSTEM 47

CHAPTER 5: VACCINES: ELEGANT, POWERFUL SIMPLICITY 67

CHAPTER 1

Disease and Therapeutic Areas of Interest

We're living in a golden age of biopharmaceutical discovery. Medicines and treatments that seemed worthy of a science fiction novel just twenty years ago are already helping people heal or are in late-stage clinical development. These include precision bioengineered cancer treatments and genetic therapies for tough-to-treat conditions. Biopharmaceutical scientists continue to push the boundaries of the possible. At the same time, our understanding of disease keeps growing as we delve further into its cellular mechanisms. In this chapter, we'll overview some of the major therapeutic areas within biopharma. Subsequent chapters explore those therapies and the technology behind them.

Infectious Disease

Tricky Terms

Pathogens and Hosts

The term *pathogen* typically refers to an infectious microscopic entity such as a virus, bacterium, or fungus. A *host* is any organism the pathogen infects. This book primarily discusses human pathogens. Nonetheless, any living organism is fair game.

Infectious diseases—also referred to as communicable diseases—are caused by **pathogens** that travel from one person to another or are passed along by exposure to contaminated food and water. Animals can even pass along some pathogens - think bird flu. These microorganisms travel in many ways, from skin to skin, through bodily fluids or feces, and by air. Sometimes, they just wait for us on doorknobs or cell phones. Infection occurs when pathogens invade a **host**, multiply, and cause our bodies to respond. Many cold or flu symptoms actually stem from that "host reaction." Our immune system kicks in, releasing the inflammatory signaling molecules that switch on white blood cells to fend off the illness. So our noses run, our joints ache, and our heads hurt.

Cocktail Fodder

Bacteriophage

Even germs get sick. A bacteriophage, or phage, is a type of virus that infects a bacterium. The mechanism bacterium uses to defend themselves forms the basis of CRISPR genome editing, discussed in Chapter 7.

Pick a Pathogen

There are so many from which to choose:

Viruses have a very simple structure. They consist only of genetic material—DNA or RNA—surrounded by a protective protein coat called a capsid. Some viruses are a bit fancier— surrounded by a lipid (fat) membrane known as an envelope.

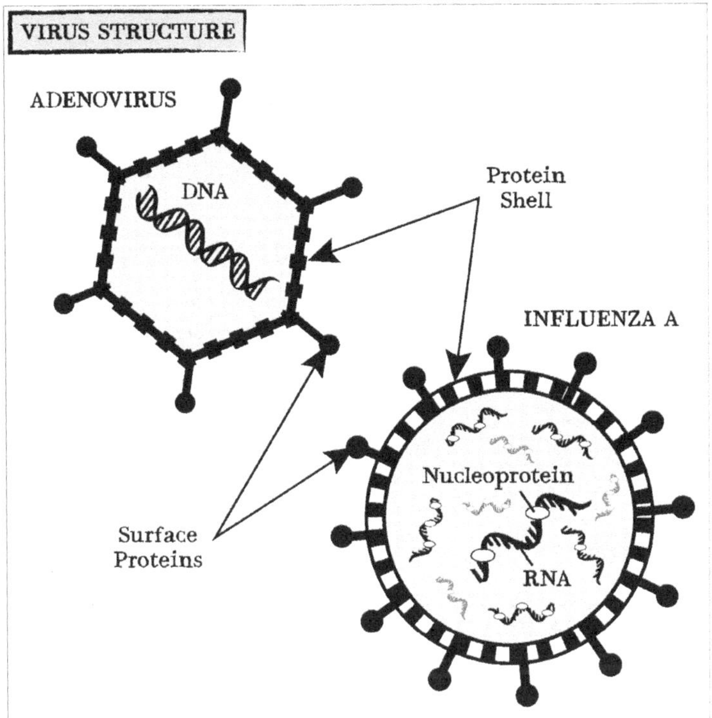

Viruses can't live on their own—they need a host cell to survive. Viruses use proteins on their surface to interact with proteins on the surface of a host's target tissue. These target-tissue proteins are referred to as **receptors**. After binding the host's target tissue receptor, the virus enters its cells. For example, a respiratory virus infects lung tissue because its surface proteins match receptor proteins on the host's lung cells—the way a key fits into a lock.

Inside a cell, a virus hijacks its machinery to replicate its genetic material and make more viral proteins. The cell is directed to assemble a new virus that spread the illness. Viruses make people sick in two ways: by directly damaging infected cells and/or by triggering an inflammatory response, resulting in symptoms that can make us miserable.

GOING FURTHER: Human immunodeficiency virus (HIV)

Human immunodeficiency virus (HIV)—which causes AIDS— disrupts a person's immune system by infecting critical white blood cells called helper T-cells. The virus kills them directly. Helper T-cells are also called CD4+ cells, because of the protein CD4 on their surface. The phrase "CD4 count" describes the number of a patient's healthy helper T-cells. It indicates the progression of HIV infection. The normal range falls between 500 and 1500. A CD4 count below 200 means a diagnosis of AIDS. Usually, the CD4 count increases if antiviral medication successfully controls the virus.

Bacteria

Bacteria are single-celled microorganisms. They can survive outside a host. The world holds trillions of different strains of bacteria. The vast majority don't cause disease. In fact, it's estimated that only about 100 species can harm us. The rest are either neutral or beneficial. Chapter 8, The Microbiome, goes into detail about "good" bacteria.

Pathogenic bacteria cause illnesses such as strep throat, tuberculosis, staph infections, and urinary tract and bloodstream infections. They make trouble in myriad ways:

- Bacteria attach to host cells, to absorb nutrients and to produce waste products, some of which are harmful. For example, the bacteria that live on our teeth consume sugar for energy and secrete lactic acid as waste. The acid breaks down tooth enamel, leading to cavities and a trip to the dentist.

- Many bacteria secrete toxins. *C. tetani,* for instance, produce toxins that interfere with regular muscle contractions and lead to spasms. These give the potentially fatal tetanus infection its more familiar name—lockjaw.
- In other cases, our body's reaction to the bacterial infection causes the symptoms themselves. For example, lungs infected with *S. pneumoniae* induce a strong inflammatory response, resulting in the build-up of pus—essentially globs of dead white blood cells. Too much pus weakens the lungs, leading to serious illness or death.

Cocktail Fodder

Surrounded

Do you ever get the feeling we're not alone? Microbiologists at the University of Georgia have estimated that more than a *nonillion*—10^{30}—bacteria live on earth. Compare that to fewer than ten billion—10^{10}—humans. It gives one pause.

GOING FURTHER: Antibiotics

Antibiotics can kill bacteria, but they're totally useless against viruses. Their structure and lifecycles are too different. For example, bacteria have their own ribosomes, which make the proteins necessary for them to thrive. Some antibiotics work by inhibiting a bacterial ribosome. Viruses don't have ribosomes; the same drugs would have no effect on them. This example illustrates why developing anti-viral medicines poses such a challenge. Viruses rely on their hosts for most of their lifecycle. Scientists must develop a drug that take out viruses without harming us—the bodies that sustain them.

Fungi

Fungi come in different shapes and sizes, from single-celled yeast to a multicellular mushroom that measures over two miles across. These organisms reproduce by forming spores. Just as we're more like bacteria than viruses, we're more closely related to fungi than bacteria. That's because both of us are **eukaryotes**: our cells feature a nucleus—an internal membrane-bound compartment that houses DNA. That's in contrast to bacteria, which are **prokaryotes**—single-celled organisms with free-floating DNA.

Like bacteria, fungi are widespread. A few make us sick. *Aspergillus* can trigger lung infection. Overproduction of *Candidiasis* causes pain and discomfort. In our mouths, the infection is called thrush. In vaginas, it's a yeast infection—the bane of many women. (Although rare, men can get yeast infections too.) Some fungi infect skin or nails, causing irritations such as ringworm or jock itch. Scientists collectively refer to these microorganisms as dermatophytes.

Cocktail Fodder

Medicine from Fungi

Some toxins from fungi kill bacteria without damaging human health. Fungi have served as an essential source of antibiotics since the discovery of penicillin from mold in 1928. Today, companies grow thousands of liters of fungi at a time to produce penicillin, cyclosporin, and other antibiotics. Cholesterol-lowering statins were also originally derived from the fungus *Aspergillus terreus*.

Genetic Disease

Genetic diseases stem from abnormalities in the genome. In the section below, we briefly touch on the different types. Our first book, *Biotech Primer One: The Science Driving Biopharma Explained,* describes the molecular mechanisms behind these diseases in more detail.

A monogenic disease stems from a defect in a single gene. Monogenic diseases are often referred to as inherited. Many rare diseases are monogenic. For example, cystic fibrosis comes from a defect in the gene that codes for a protein that helps regulate fluids in the lungs. Gaucher's disease arises from a mutation in the gene that codes for an enzyme that helps digest large molecules within cells. Finally, severe combined immunodeficiency arises from a malfunctioning gene responsible for a protein critical to the immune system.

Polygenic diseases are far more common. They originate in the interactions of many genes. Such conditions include cancer, heart disease, Alzheimer's disease, and Parkinson's disease. Polygenic diseases often have **susceptibility genes** linked to them. These increase the chances of developing a disease but don't absolutely predict it. A person's medical destiny depends on the interplay of their other genes, lifestyle, and environment.

One well-known example of the role of susceptibility genes in human health is the association of breast cancer with the BRCA1 and BRCA2 genes. Mutations in these genes make someone more likely to get cancer, but don't cause cancer. BRCA1 and BRCA2 code for proteins that repair damage to DNA. If these proteins are compromised, they're less likely to fix other mutations that actually cause the disease. Women with harmful mutations in either BRCA1 or BRCA2 have a risk of breast cancer that is about five times the normal risk.

Our genes are arranged on long strands of DNA called chromosomes. **Chromosomal abnormalities** include deviations in chromosome number or structure. Human cells have 23 pairs of chromosomes for a total of 46. Variations from this number can cause disease. For example, trisomy 23—three copies of chromosome 23 instead of the usual two—results in Down syndrome. Irregularities in chromosome number may also trigger miscarriage.

Chromosomal translocations occur when a segment of one chromosome gets transferred to another chromosome or a new spot on the same chromosome. Some translocations are harmless; others are linked to cancer or infertility.

GOING FURTHER: Chromosomes

Abnormalities in chromosomal number typically arise in *meiosis,* the cell division unique to the production of sperm and egg cells. These cells copy their genetic material once and divide twice. This ensures that each new egg or sperm contains exactly half a person's genetic material. When a sperm cell fertilizes an egg, it restores the correct chromosomal number. Sometimes, however, a mistake occurs during meiosis and an egg or sperm cell ends up with two copies of a chromosome instead of one copy. The resulting zygote (fertilized egg) will have an extra chromosome.

Mitochondrial DNA

The cell nucleus contains most of our DNA. However, a small amount—exactly 37 genes—is found in mitochondria, the organelle that converts sugar to chemical energy in cells. Mutations in these mitochondrial genes are associated with rare diseases such as myoclonic epilepsy and Leber's hereditary optic atrophy, a type of retinal degeneration. Mitochondrial DNA is passed down via egg cells, thus only from our mothers. Not all mitochondrial disease originates in defective mitochondrial DNA. Mutations in nuclear genes that code for proteins in mitochondria can also cause disease.

Cocktail Fodder

Mommy and Daddy, plus Mommy?
Many couples struggle to conceive. If a mother knows that she carries a gene responsible for mitochondrial disease, the situation gets more complicated still. Now, a novel procedure provides hope to prospective parents. "Three-parent *in vitro* fertilization (TPIVF)" aims to prevent the transmission of genetic disease in mitochondrial DNA.

During TPIVF, doctors remove the nuclear DNA from the affected mother's egg. The doctor then replaces the donor egg's nuclear DNA with the mother's nuclear DAN. However, the donor egg's undamaged mitochondrial DNA remains intact. Doctors implant the new and improved egg into the prospective mother. If all goes well, the procedure ultimately results in a healthy baby. The newborn has DNA from three parents—two women and one man.

Oncology

At its essence, cancer is uncontrolled cell growth. Healthy cells exert tight controls on cell division. They divide only in response to outside signals for cellular growth and division. Normal cells also react to signals to *stop* dividing. For example, most cells exhibit contact inhibition. Touch a neighbor, and division ceases. Cancer cells have lost many of their reproductive checks and just keep multiplying.

What causes this runaway growth? Typically, it results from mutations in one or more of the genes that code for proteins that regulate cell division. For example, a growth factor receptor protein tells a cell to grow and divide in response to chemical signals called, unsurprisingly, growth factors. With a mutation, the growth factor protein could direct a cell to divide *without* growth factors. Other proteins react to signals for a cell to stop dividing. Mutations here could make the cell ignore the "stop" signals. In most cases, a malignant tumor contains errors in several genes. A big part of oncology research involves identifying mutations that cause cancer and designing therapeutics that address them.

Autoimmune Disease

An **autoimmune disease** occurs when the immune system mistakes part of the body for a threat. This causes white blood cells to attack that tissue. Symptoms vary by disease. For example, people with rheumatoid arthritis suffer from painful, stiff, and swollen joints because their immune system damages them. People with Crohn's disease suffer abdominal pain, diarrhea, and weight loss as their immune system targets the gastrointestinal tract. Misfiring white blood cells also release inflammatory signals, leading to chronic inflammation.

The roots of autoimmune disorders remain unknown. However, effective treatments, such as Humira, alleviate symptoms. They work by neutralizing some of the signals that over-activate immune response.

Neurological Disease

Neurological diseases affect the central nervous system (CNS)—the brain and the spinal cord. The CNS sends and receives information from our peripheral nervous system, the vast network of nerves that feed into our every part. Its signals enable voluntary and involuntary movement. They also allow our brain to process and interpret sensory information from the spinal cord.

Specialized cells called *neurons* make up the CNS. They send and receive signals electrochemically. A *neurotransmitter*—a chemical message—is converted into an electrical signal within the neuron. When this charge reaches the edge of the neuron, the *dendrite*, it causes ion channels in the cell membrane to open. The resulting gap allows positively charged sodium ions to enter, sending the electrical signal through the body of the neuron. The charge leaves through the cell's opposite side, the *axon*. This release causes other neurotransmitters to activate neighboring neurons. Different neurons send and receive various neurotransmitters.

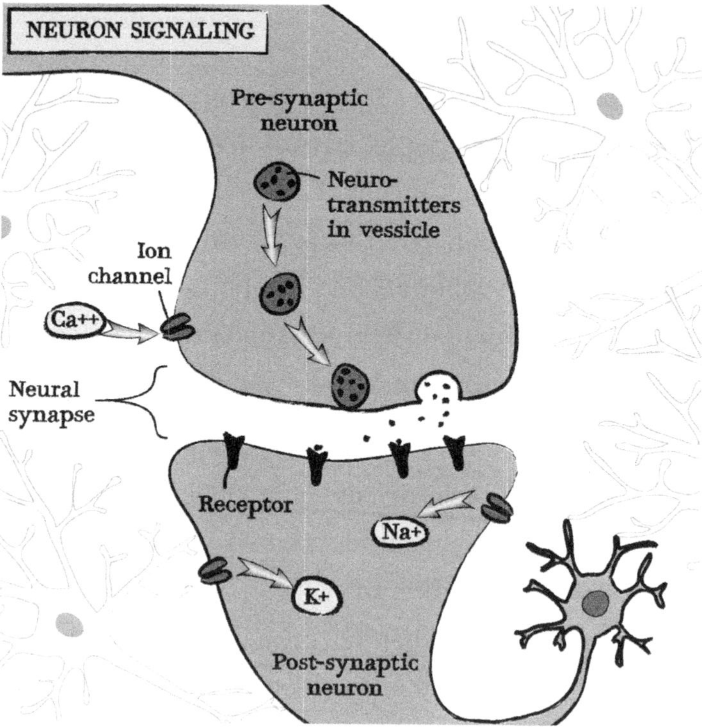

Billions of neurons within the central nervous system communicate through more than a hundred neurotransmitters, such as serotonin or dopamine. This neuron/neurotransmitter duo regulates just about every aspect of the human body: movement, hunger, body temperature, emotion, even wakefulness.

Not surprisingly, we don't entirely understand such a complex system or the conditions that afflict it. The CNS can fall prey to infectious, genetic, or neurodegenerative diseases, cancer, stroke, or traumatic injury. Four of the top neurological research priorities within biopharma today are Huntington's disease (HD), Parkinson's disease (PD), multiple sclerosis (MS), and Alzheimer's disease (AD).

Huntington's Disease (HD)

This is a neurodegenerative disorder—in which neurons progressively lose structure and function. As the disease continues and more neurons are damaged and die, symptoms worsen. Early-stage patients experience subtle involuntary movements and mood disturbances like depression and anxiety. Later, sufferers lose the ability to walk, speak, and even swallow. Life expectancy is twenty years after its initial symptoms. Ninety percent of HD cases affect adults between the ages of 30 and 50; juvenile-onset occurs in the remaining ten percent.

HD is monogenic, caused by a mutation in the huntingtin gene. The disease is dominant. If someone's parent suffered from HD, he has a 50 percent chance of developing it too. Despite knowing the genetic basis of HD, scientists don't yet fully understand its mechanism. Recent advances in gene therapy and genome editing (fixing defective gene copies with functional ones) offer hope for a cure. Another approach under investigation involves swapping dead or damaged neurons with replacements derived from stem cells.

Parkinson's Disease (PD)

Parkinson's disease is also neurodegenerative. In particular, PD patients experience reduced activity and the death of neurons that secrete the neurotransmitter dopamine. Typical symptoms include tremors, slowness of movement, rigidity, and impaired cognitive function.

PD primarily affects people over 60. As the Baby Boomer population ages, public health experts expect significant increases in PD.

There is no specific gene associated with PD. In fact, most cases are classified as "sporadic," arising without a genetic association or apparent cause. Current research involves teasing out genes that may indicate susceptibility. For now, there's no cure. However, drugs that mimic dopamine's effects can help manage symptoms. Ongoing research also involves using stem cell-derived neuronal cells to replace dead or dying neurons.

Multiple Sclerosis (MS)

This autoimmune disease causes the patient's immune system to attack and destroy the protective insulating layer that surrounds the axon section of the neuron. This interferes with the transmission of electrical signals from one neuron to the next.

MS typically strikes young adults. Early signs of the disease depend on what part of the nervous system it hits. They can include problems such as difficulty walking, numbness or tingling sensations, and blurred vision. Initial symptoms often come and go and may not reoccur for years. MS is chronic and progressive, gradually worsening over decades. Patients often end up in wheelchairs.

MS has no clear-cut genetic cause, although researchers have identified some susceptibility genes. The disease appears to have many conflating factors. It also occurs much more in women than in men by a ratio of three to one. This disparity suggests that female hormones play a role. Geography is also a factor—the farther north one travels from the equator, the greater the incidence of the disease. Because sunlight is required to produce vitamin D, this has led to a proposed connection between a lack of vitamin D and MS. Researchers at Johns Hopkins University and the Cleveland Clinic are currently investigating this connection to determine if vitamin D supplementation may help protect susceptible individuals. Finally, viral infections may trigger MS as well. The development of the disease likely results from a combination of genetic and environmental factors.

Although there's no cure, many drugs can slow the progression of MS by blunting the immune system's attack on the CNS.

Alzheimer's Disease (AD)

Alzheimer's disease accounts for approximately seventy percent of dementia cases. Like Huntington's disease and Parkinson's disease, it's neurodegenerative. Neurons in the brain's hippocampus, the region associated with memory formation, are among the first afflicted. By 2025, the number of people aged 65 and older with Alzheimer's disease is projected to surpass seven million. That's a forty percent increase from the more than five million affected in 2015 (Alzheimer's Association).

Alzheimer's disease is associated with the build-up of amyloid-beta (Aβ) plaques in patients' brains. They derive from the cleavage of the amyloid precursor protein, which may play a role in forming synapses. Individual Aβ molecules clump together to create these plaques.

Until very recently, we didn't know how Aβ plaques might cause Alzheimer's. Researchers at Stanford School of Medicine have demonstrated that Aβ binds to a receptor on nerve cells that impairs synapse function. This finding suggests a potential drug target. Disrupting this interaction could preserve nerve cell function.

No cure exists for AD at this time. However, a number of companies are working on treatments. A few have already made it to clinical trials. We'll look at these in more detail in Chapter 9.

In this chapter, we've looked at some major therapeutic areas of interest to biotech companies today. Next, we review how our body fights disease.

CHAPTER 2

How Your Body Fights Disease

The immune system is the body's primary line of defense against infection and cancer. Understanding its function is crucial to developing new drugs and therapies for two reasons. First, medications often have to work with the immune system to produce optimal effects; many new biotech drugs rely on the mechanics of immune function. Second, a significant number of illnesses are either diseases of the immune system or arise from a malfunctioning immune system. With all that in mind, we'll take a thorough look at our immune system and how biotech is wielding its knowledge of its mechanisms in pursuit of human health.

Many Cells, One Goal

All immune system cells come from one identical population, **hematopoietic stem cells (HSCs)**. HSCs reside in the bone marrow. There, they divide to produce identical stem cells or **progenitor cells**. HSC also generate niche cells, particularly **stromal cells** and **osteoblasts**, which help regulate the HSCs and their differentiation. Hematopoiesis, the process by which blood cells are formed, takes place mainly in the bone marrow. It unfolds in an orderly sequence via progenitor cell lines and ultimately gives rise to all types of blood cells. Familiar immune cells include red blood cells (**erythrocytes**), white blood cells (**leukocytes**), and **platelets**. Specific sub-classes of white blood cells, including **B-** and **T- cell lymphocytes**, **neutrophils**, and **macrophages**, play a fundamental role in immunity.

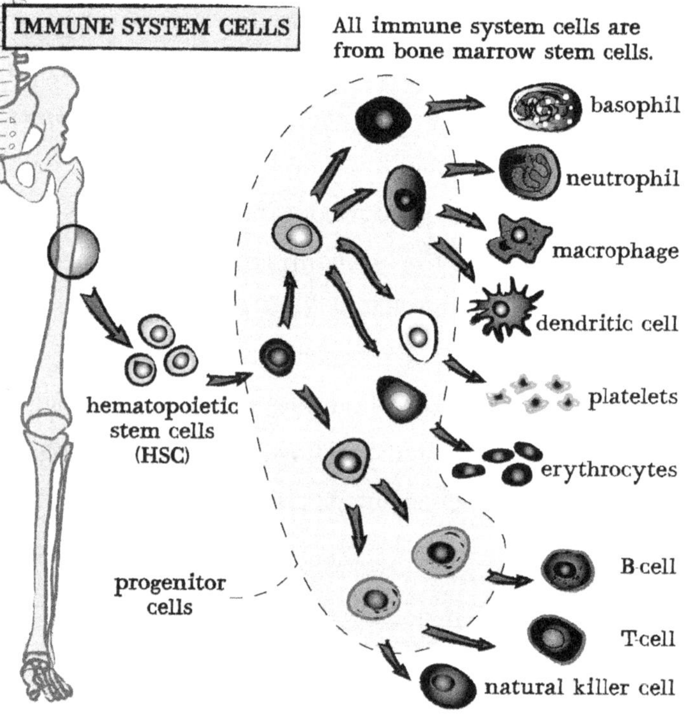

Red blood cells carry many copies of hemoglobin. This key protein binds and transports oxygen throughout the body. The reaction between oxygen and the iron-containing heme group in hemoglobin gives red blood cells their hallmark color.

Platelets, or **thrombocytes**, are small cells without nuclei. They circulate in mammalian blood and are involved in **hemostasis** or blood clotting. If an organism is low on platelets, it may be subject to excessive bleeding. *Too many* though, can cause blood clots, increasing the risk of heart attacks or stroke.

Lymphocytes, also known as B- and T-cells, are the heavy hitters. B-cells and T-cells recognize and destroy foreign cells and viruses. B-cells design, produce and secrete antibodies that tag foreign antigens—substances, usually proteins, that induce an immune response—for annihilation. The word "antigen" is a mashup of "antibody" and "generator." Antigens include proteins in bacterial cell walls and the protective protein coat around viruses.

T-cells handle the destruction of foreign or dysfunctional cells. Think infection or cancer. **Helper T-cells** release chemical messages known as **cytokines** to recruit other immune cells, such as cytotoxic T-cells and macrophages, to destroy infected or foreign cells. **Cytotoxic T-cells** release toxins that kill them directly. The work of B-cell and T-cell is complex. We'll revisit them further in the chapter.

Phagocytes ingest (eat!) microorganisms. **Macrophages** engulf and digest bacteria and other aliens. **Granulocytes**, such as **basophils** and **eosinophils**, perform a similar task by secreting chemical granules that digest invaders. **Neutrophils** are twofers, granulocytes, *and* phagocytes in

Cocktail Fodder

Skin is an organ

It's sometimes hard to remember that skin is an organ, let alone part of the immune system. And it's even our largest organ, no matter what our hearts or brains say. Our skin weighs a hefty ten or so pounds in a typical adult.

one miniscule cell! They swallow and digest any interlopers with the help of granules they release.

Cocktail Fodder

Mucous
Here's a metaphor you don't see every day: Mucous is flypaper for bacteria and viruses. It's wet and sticky and catches little pests that you prefer to keep outside yourself. Kleenex® anyone?

The Outer Perimeter: Non-Specific Immune Response

Our bodies don't rely solely on internal means to stave off disease and infection. The first, external defenses against **pathogens** such as viruses, bacteria, and fungi are quite familiar and, well, "earthy." They include our skin, mucous membranes lining the gut and the lungs, and nostril hairs. Bodily secretions get in on the action, too: tears, saliva, post-nasal "liquids," sweat, and strong stomach acids. Sneezing and coughing also help fend off nasty microbes. As you can see, we are, in fact, total busy bodies.

If pathogens make it past our outsides, they hit our **non-specific immune response**. This defends us from infection by pathogens *generally*. Key players here are the tiny, powerful macrophages, neutrophils, and eosinophils.

GOING FURTHER: Inflammation

Inflammation is a response to a wound, irritation, or infection. Ideally, it minimizes invasion by bacteria and other pathogens. The response also involves sending a host of immune cells to the site of inflammation, including granulocytes and macrophages.

Consider the good old-fashioned cut. When you slice your finger, bacteria pour into the wound. The invasion prompts the injured cells to begin unleashing **cytokines** and

chemokines. These signaling molecules attract immune cells. The chemicals activate immature immune cells, **monocytes**, to leave the bloodstream and enter the wound area. There, they change into macrophages and start eating the bacteria and damaged cells. When the macrophages get the situation under control, the injured cells stop calling for help. Tah dah—the inflammation subsides. By the way, the telltale pus that accumulates in or on an infection consists of dead and decaying cells.

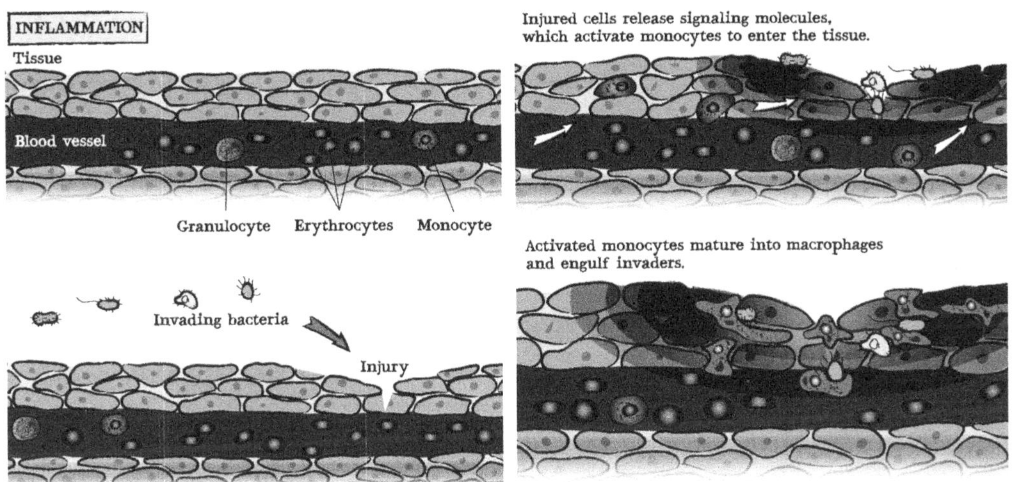

Redness, swelling, pain, loss of function, and a localized feeling of warmth are classic signs of acute inflammation. Many of the body's first responses to the trauma are mediated by the outflux of signaling molecules from white blood cells. For instance, **prostaglandins** trigger **vasodilation**, which is a widening of blood vessels that allows increased blood flow. These signaling molecules are short-lived, which helps quickly resolve the inflammatory response.

Resolve? The term **resolution** designates the body's return to a normal non-inflammatory state. Resolution is an active process, like inflammation. When resolution fails to occur, chronic inflammation may result. It's not associated with the traditional marks of acute inflammation. But it also involves destruction or damage to body tissues, often by immune cells that rallied to the tissues to fight infection. Chronic inflammation also results from unresolved autoimmune reactions that may develop into serious autoimmune diseases.

Call in the Reinforcements: Specific Immune Response

Our non-specific defenses or surface barriers, such as skin, don't always succeed. Then it's time for a **specific** or **adaptive immune response** to kick in. Here's where the body's "big guns," the B-and T-cell lymphocytes come to the rescue.

Lymphocytes are highly specialized and patrol continually throughout the circulatory and **lymph system** for foreign or "non-self" cells. They're looking for antigens that are unknown, usually foreign proteins. During a medical exam, the doctor feels lymph nodes in your neck and armpit. Swelling could indicate that you're producing loads of white blood cells to knock out an infection.

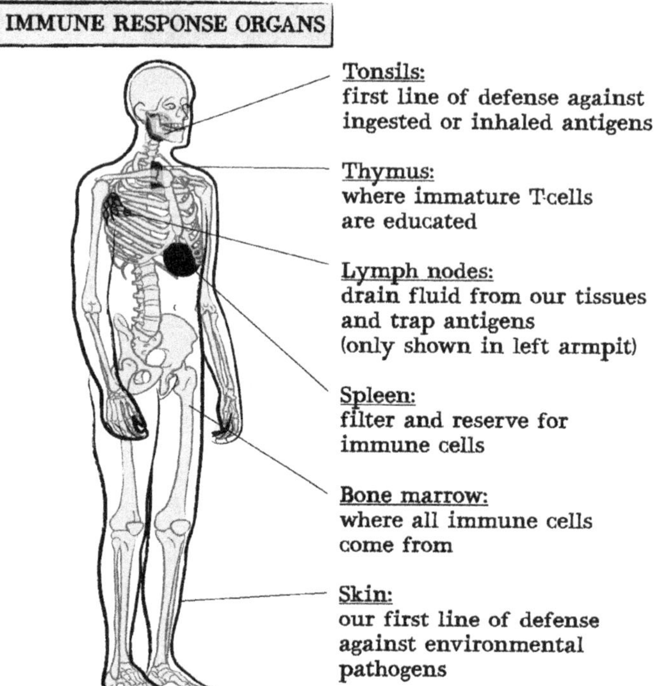

Although the **innate** (non-specific) and **adaptive** (specific) immune systems both protect us, they differ substantially. The adaptive immune system needs time to react to an invading organism. In contrast, the innate immune system includes mostly ever-present and ever-ready defense mechanisms. Second, the adaptive immune system is antigen-specific; it reacts only to the particular trigger. Conversely, the innate system reacts equally well to a slew of organisms. Finally, the adaptive immune system demonstrates **immunological memory**. It "remembers" invading organisms and reacts faster upon subsequent exposure.

Mr. (and Ms.) T

Though they originate in bone marrow, T-cells mature into **T-helper cells** or **cytotoxic T-cells** (aka "Killer T-cells") in the **thymus**. Hence, the *T*. These cells have unique **T-cell receptors (TCR)** on their surface. Both types activate upon recognizing foreign antigens.

Helper T-cells recognize antigens displayed on other immune cells (such as macrophages, dendritic cells, or B-cells) that have engulfed the antigen and are now called **antigen-presenting cells (APC)**. They coordinate the overall immune response by secreting cytokines, including **interleukins** and **interferons**. Cytokines activate and direct other immune cells, such as cytotoxic T-cells and macrophages. For instance, interleukin 2 stimulates T-cells to proliferate. It's released by helper T-cells during a viral infection. Helper T-cells can also activate B-cells by the cytokines released, which produce antibodies to help clear the pathogen. These helper cells can't kill infected invaders. They recruit other immune cells to do the dirty work.

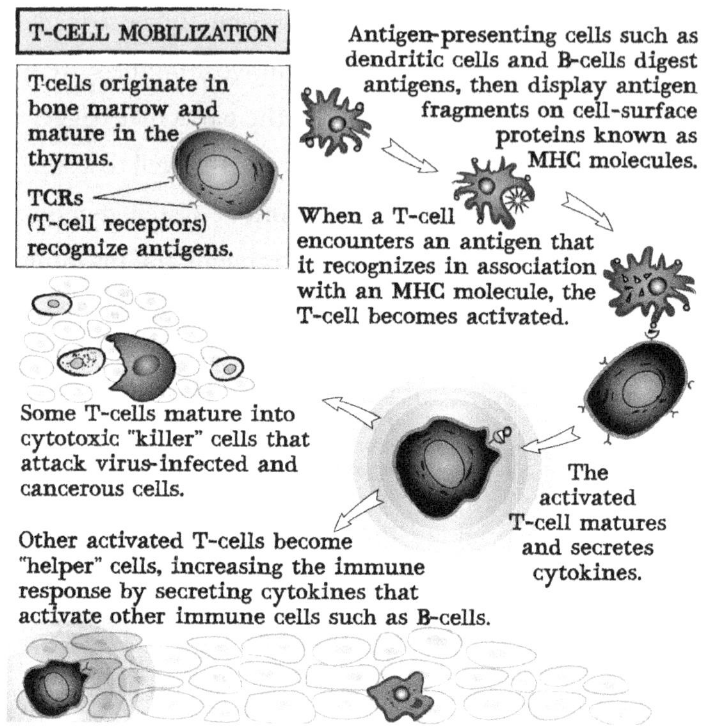

Cytotoxic T-cells kill infected host and damaged or dysfunctional cells, including those that are cancerous. Somewhat ironically, they're also responsible for rejecting tissue and organ grafts, wreaking havoc in seriously ill patients. Cytotoxic T-cells release **cytotoxins**, which create holes and a reaction inside the targeted cells that lead to their death.

The Regula-T-ors

The third type of T-cell, **regulatory T-cells (Tregs)**, actually suppresses the immune system. They prevent our own immune cells from attacking our bodies. Like other T-cells, Tregs switch on when their receptors recognize a particular antigen. The bit of the antigen to which the Treg receptor responds is a **Tregitope**. Scientists haven't totally figured how Treg suppression works. They *do* know the effect comes partially from the release of anti-inflammatory cytokines. Defective or insufficient Tregs can spur killer T-cells, macrophages, and other attack cells into overdrive, causing severe inflammatory disease.

GOING FURTHER: Immune System Checkpoints

A healthy immune system is finely calibrated between over and under activity. A strong, rapid reaction makes it possible for us to fight infection. If our immune system turns on our own bodies, however, serious disease results. For example, when our immune system attacks pancreatic insulin-producing cells, the life-threatening result is Type 1 diabetes.

Immune system checkpoints are one of the mechanisms that prevent cytotoxic T-cells from attacking our healthy bodies. These are proteins on the surface of cytotoxic T-cells that send an inhibitory signal to T-cells when they encounter proteins on the surface of healthy cells. The checkpoint quickly shuts down the killing power of the cytotoxic T-cells, leaving our cells unharmed. One of the best characterized checkpoint proteins is PD-1. When it runs into the protein PD-L1 on the surface of normal, healthy cells, its inhibitory activity kicks in, telling cytotoxic T-cell to lay off.

Cocktail Fodder

Nobel Prize
The 2018 Nobel Prize for Physiology or Medicine went to **James Allison** and **Tasuku Honjo** for their discovery of immune checkpoints in the early 1990s.

Tricky Terms

Interleukins
Interleukins got their name because they're messengers for white blood cells. *Inter* means "between." *Leukin* describes white cells.

Tricky Terms

Interferons
Though they perform many functions, interferons were named for their ability to *interfere* with viral replication.

B-Cells

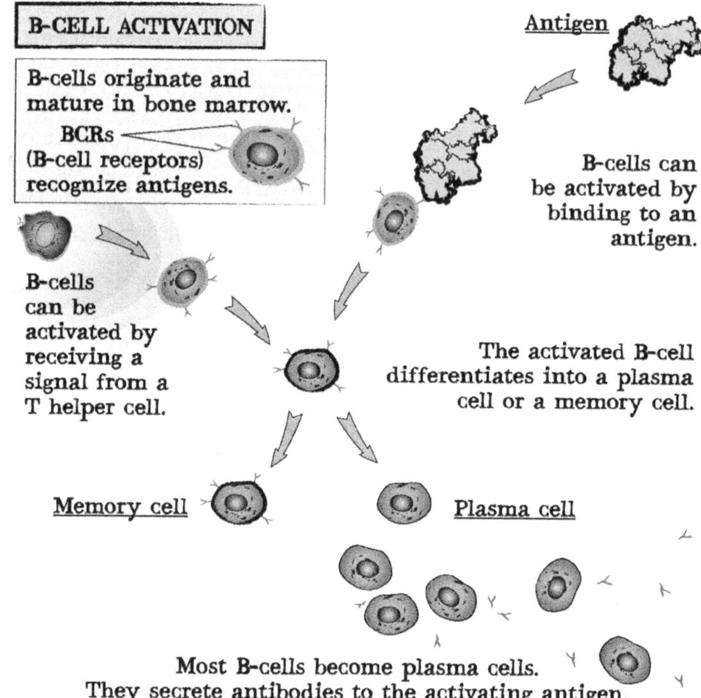

Like T-cells, B-cells originate in the bone marrow. They mature there too, hence the "B" designation. B-cells have specific **B-cell receptors (BCRs)** on their surface to recognize antigens. B-cells can become activated either by binding to an antigen or by receiving a signal from a T-helper cell. This latter is called **T-cell-dependent activation**.

When B-cells are activated, they differentiate into **plasma cells** or **memory cells**. Most B-cells become plasma cells that produce and secrete antibodies. A particular antibody is simply a soluble form of the receptor that was activated. In other words, each B-cell secretes antibodies that target the same antigen that initially activated it.

Other B-cells become long-living memory cells, which circulate and patrol for a future invasion of that particular antigen. This allows the body to respond quickly if it gets reinfected with the same pathogen.

Cocktail Fodder

B-cells

You are born with an abundance of different B-cells, each of which recognizes a different antigen. Somewhere in your crowded bone marrow lies a B-cell that will recognize a new strain of influence decades from now and launch an immune response. Talk about planning ahead!

Antibodies

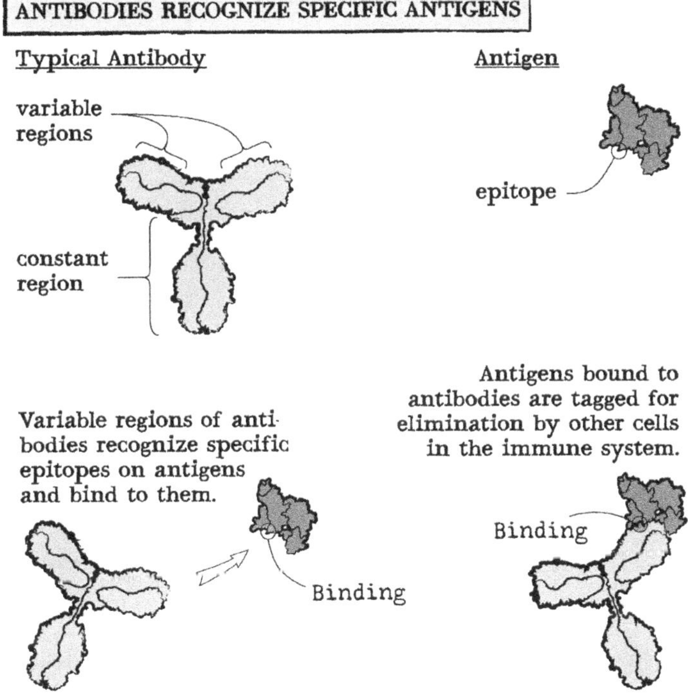

Each plasma cell is essentially a factory that produces one antibody tailored in response to a specific antigen. Antibodies are Y-shaped, with a variable region at the top of the "Y" branch that enables it to recognize and bind to a specific portion of its antigen—the **epitope**. Each antibody has a unique variable region. That explains the existence of

Cocktail Fodder

Antibodies and Macrophages

An activated B-cell turns out up to 50,000 antibody molecules each minute, and a macrophage can consume as many as a half dozen bacteria at a time! The word macrophage literally means "big eater." No wonder we're tired when we're sick. Our bodies work very hard to battle infection.

millions of different antibodies, each recognizing a different epitope. The interaction between antibody and epitope is highly specific, much like a lock and key.

Antibodies attach to foreign antigens and tag them for elimination by other cells of the immune system. The consequences of the secreted antibodies binding to their antigens can vary. Binding and covering antigens on the cell surface of an invading microbe can cause the microbes to clump together and become incapacitated. Or it can signal macrophages to come eat the invaders, *or* the antibodies may bind to a toxin's active site and neutralize it.

The activation of the B-cell into plasma and memory cells is called the **primary response**, which lasts for several days or weeks. The concentration of antibody decreases as the plasma cells stop secreting them. Once the infection is eradicated, plasma cells die, but memory cells remain in the body.

If the same pathogen causes another infection, then the immune system encounters antigens it already "knows." This triggers the **secondary response** faster because the existing population of memory cells can produce many plasma cells with the appropriate antibody. These destroy the pathogen before it causes any symptoms.

INDUSTRY NOTE: Autoimmune Disorders

Autoimmune disorders can sometimes be managed with biologic drugs designed either to interfere with activating cytokines or to supplement inhibitory cytokines.

Biotech Applications of Antibodies

The ability of antibodies to recognize and bind to one specific antigen has been exploited to a great advantage in the biotech industry. Antibodies have been used to great success in both therapeutic and research applications. We'll discuss some of these applications later in this chapter. First, though, how do scientists even obtain antibodies?

Cocktail Fodder

Antibodies

Scientists estimate that we generate about ten *billion* different antibodies, each capable of binding a distinct epitope of an antigen.

Busy Bunny B-Cells

Scientists make **antiserum** by injecting an animal, usually a rabbit, sheep, or goat, with the protein they want to make antibodies against. The B-cells recognize the protein as foreign and produce antibodies that bind to multiple epitopes on that protein. Each successive immunization produces increasingly more antibodies thanks to the animal's memory B-cells. These are **polyclonal antibodies**.

This kind of antibody derives from different B-cell lines. They're a mixture of antibodies secreted against a specific antigen, each recognizing a different epitope (each antigen contains many different epitopes). Polyclonal antibodies are used as diagnostics and as research tools. However, therapeutic antibodies need to be more specific, recognizing only one epitope hence the name **monoclonal antibody**. This ensures a consistent, predictable therapeutic effect.

Attack of the Clones

This specificity problem, having an antibody recognize just one epitope, was overcome with **monoclonal antibodies** in the mid-1970s. Kohler and Milstein developed this technology and received a Nobel Prize in Physiology or Medicine for it in 1984.

Making monoclonal antibodies relies on the **clonal selection** of B-cells, meaning researchers isolate one B-cell (also called a clone) to produce identical antibodies that all bind to the same epitope. Monoclonal antibodies are most often made from mouse B-cells. First, researchers immunize the mouse with the protein antigen. Once the mouse has generated an immune response, scientists remove its spleen to harvest the cells.

B-cells don't survive in cell culture, so scientists "immortalize" them. They fuse B-cells to **myeloma** cancer cells. If you recall, malignancies are notoriously robust. The resulting hybrid cell is appropriately called a **hybridoma**.

Hybridomas possess the antibody-producing capability of B-cells and the immortality of the myeloma cells. Once researchers isolate hybridoma cells, each gets screened to determine if it's yielding antibodies with the desired characteristics.

Humanizing Mouse Antibodies

Though monoclonal antibody (mAb) technology was invented in the mid-1970s, its promise in human therapy wasn't realized for almost 20 years. Monoclonal antibodies initially failed because our immune system treated the mouse antibodies as foreign. The resulting **human anti-mouse antibody (HAMA)** response decreased monoclonal antibodies to ineffective levels. It also posed the threat of serious allergic reactions. Monoclonals' current success in immunotherapy comes from our capacity to make mouse antibodies more human. Once scientists isolated hybridoma clones expressing monoclonal antibodies, they learned to manipulate the immunoglobin genes encoding those antibodies through recombinant DNA technology. Remember the variability of antibody structure? By swapping out the variable region from the human immunoglobulin gene for the corresponding variable regions from monoclonal antibody-encoding mouse genes, and *then* introducing the hybrid gene into mammalian cells to be expressed, researchers created monoclonal antibodies with a mouse antibody-antigen specificity. But because these are still mostly human, these antibodies are called **humanized**. It's now also possible to generate fully human monoclonal antibodies from transgenic mice engineered to express the human immunoglobulin gene.

Cocktail Fodder

White Blood Cells
White blood cells account for only one percent of the cells in the five liters of blood in an adult's body. But don't worry, there are more than enough white blood cells to get the job done: In each liter of blood, you have 5,000,000 white blood cells.

HUMANIZING MOUSE ANTIBODIES

Mouse DNA expresses mouse antibodies.

Mouse DNA

Mouse antibody

When injected into a human, mouse antibodies are attacked by human anti-mouse antibodies.

HAMA

Recombinant immunoglobulin genes, expressed in mammalian cells, produce humanized mouse antibodies.

Mostly human DNA

Humanized mouse antibody

INDUSTRY NOTE: Human Antibodies

The latest generation of monoclonal antibodies are made using mice that have been genetically engineered to contain human antibody genes. When these mice are injected with antigen as described above, the antibodies produced are fully human. There's no need to manipulate them further to ditch the mouse DNA.

Researchers have created monoclonal antibodies that treat a number of diseases, including cancer, cardiovascular disease, inflammatory diseases, macular degeneration, transplant rejection, multiple sclerosis, and viral infection. This type of treatment is called monoclonal antibody therapy.

In the next chapter, we see how new drugs are categorized.

Cocktail Fodder

Starfish

Starfish have an innate immune system that is similar to our own.

CHAPTER 3

Therapies: How Biopharma Fights Disease

Medicine comes in many different shapes and sizes—from Alka Seltzer to the flu vaccine to nucleic acid-based drugs that bind to messenger RNA and actually turn off a problematic gene. Currently, there are over 20,000 prescription medicines on the market. This chapter overviews the different types of therapeutics. Subsequent chapters will look at immunotherapies, vaccines, gene therapies, and nucleic acid drugs in depth.

Broadly speaking, therapies are separated into **small molecule drugs** and **biologic** or **large molecule drugs**. These, in turn, fall into several sub-categories.

Small Molecule Drugs

Small molecule drugs are what we think of as tradi-tional" pharmaceuticals. People take them as pills or swal-low, rather than inject them. Acetylsalicylic acid, otherwise known as aspirin, serves as the classic example.

Tricky Terms

Drugs Defined

Perhaps more so in biotech than in other areas, precision matters. This certainly holds true for terminology. The Food and Drug Administration (FDA) defines a **drug** as a substance:

- Recognized by an official **pharmacopoeia** or **formulary**.
- Used in the diagnosis, cure, mitigation, treatment, or prevention of disease.
- Intended to affect the body's structure or function.
- Used as a component of a medicine. (However, devices or their components, parts or accessories are *not*.)

Manufacturers generally synthesize small molecule drugs using standard chemical processes. Essentially, this means mixing together a medicine's ingredients in a specific order under a specific temperature and pressure.

These medicines are called "small molecule" because they're tiny enough to actually enter cells. This characteristic differentiates them from large molecule biologic drugs. Small molecule drugs typically consist of fewer than a hundred atoms and possess a simple chemical structure. Most medicines on the market today are small molecule, although biologics are becoming more widely available.

GOING FURTHER: Drugs from Nature

Acetylsalicylic acid has been chemically synthesized and sold as a highly effective anti-inflammatory since the late 19th century. How was it discovered? No one will ever know for sure, but clay tablets from ancient Egypt suggest healers used the leaves and bark from willow trees to treat aches and fever. Both contain high amounts of salicylic acid. Other historical references to the anti-inflammatory power of willow bark include writings by Hippocrates. By the mid-1800s, pharmacists were prescribing extracts of willow bark. Around the same time, the French chemist Henri Leroux identified salicylic acid as the magic compound.

In the late 1800s, scientists at the Bayer dye and drug company in Barmen, Germany, created a synthetic derivative of salicylic acid called *acetyl*salicylic acid. The manufactured version irritated people's stomachs much less than the natural remedy. Many other drugs that are now chemically synthesized originated from plant extracts. Some of these are the cancer medicine Taxol (Pacific yew tree), Digoxin (foxglove) for heart failure, and artemisinin (sweet wormwood) for malaria.

Biologics

Biologics also come from nature, just not from the forest, jungle, or garden. They derive from living materials, usually cells. The most common type of biologic drug is protein therapeutics. Examples include insulin, growth hormones, and monoclonal antibodies. The only way to make proteins thus far is to do it in a cell. Scientists use **genetic engineering** to transfer a gene encoding the protein into a particular kind of cells called the *production cell line*. Depending on the product, companies grow thousands of liters of these engineered cells at a time. This process is called **biomanufacturing**. To learn more about genetic engineering and biomanufacturing, please refer to our first

Tricky Terms

Biologic, Protein Therapeutic, and Large Molecule Drug
The terms biologic, protein therapeutic, and large molecule drug are interchangeable. They all refer to a protein that treats a disease or health condition. The term "large molecule drug" describes a protein therapeutic size relative to that of a small molecule drug; biologics range from fifty to a thousand times bigger than a typical small molecule medicine such as ibuprofen.

book, *Biotech Primer One: The Science Driving Biopharma Explained.*

Other Types of Biologics

The FDA classifies several other product categories as biologics. They include vaccines, gene therapies, and cell therapies. Each requires a living cell or organism to produce them. You'll find more information about them in the chapters that follow. Industry professionals generally refer to protein therapeutics as "biologics" or "biologic drugs." Others are described using the name of their specific category—e.g., vaccines or gene therapy.

Peptide Therapeutics

The FDA defines a *peptide therapeutic* as a molecular chain of forty or fewer amino acids (the building blocks of proteins.) Peptide therapeutics and small molecule drugs are similar in that they can be constructed using a peptide synthesis machine, which links amino acids in a specified order. Peptides also share key characteristics with large molecule drugs. These include sensitivity to digestive enzymes, delivery by injection, and high specificity for their target. Nonetheless, the FDA classifies peptides as small molecule drugs.

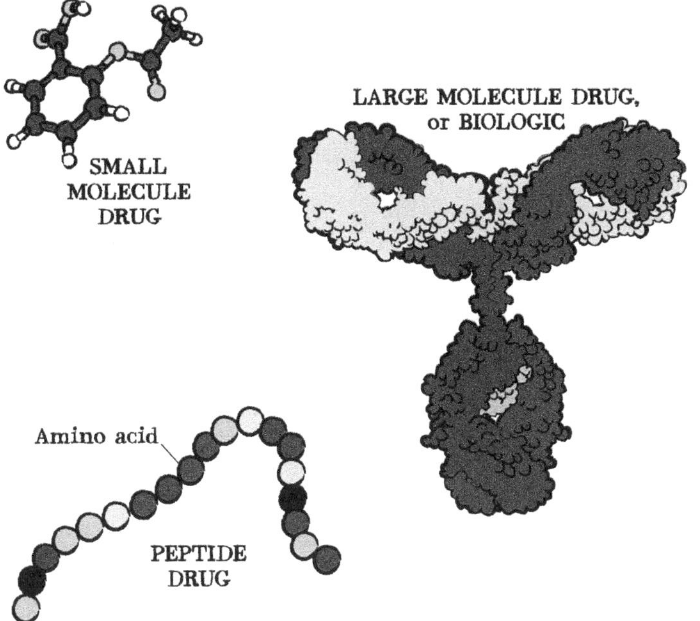

Some of the most common peptide drugs are glucagon-like peptide-1 receptor activators (GLP-1). These synthetic peptides treat type 2 diabetes by interacting with a receptor on the surface of pancreatic beta cells, which stimulates the release of insulin. GLP-1 activating peptide drugs include Byetta, Victoza, and Trulicity.

Tricky Terms

Biologics, Biosimilars, and Generic Drugs

The FDA classifies biologics that "copycat" already approved large molecule drugs or biologic medicines as **biosimilars**. The original drug is known as a **reference product** or **innovator drug**. The manufacturer of the biosimilar confirms its similarity to the reference product through laboratory and clinical testing. A **generic drug** is a small molecule drug that copies the reference product atom by atom.

Biosimilars and Generics

Why are biosimilars classified only as "similar" and not exact copies? The highly complex structure of the proteins that comprise biologic drugs makes it impossible to analyze and duplicate their precise molecular structure.

In addition, proteins are exquisitely sensitive to their environment. For example, only a small shift in pH can change their structure. It's likely that any minor changes in how the original is made—in the cell lines used to make the drug, the manufacturing environment, or in the biosimilar's final **formulation**—will cause slight structural differences between the biosimilar and the reference product.

Since it's technically impossible to detect all potential differences, the FDA requires biosimilar manufacturers to conduct at least some clinical trials to confirm their product's safety and effectiveness.

The makers of generics, in contrast, don't need to run clinical trials. Instead, they can rely on the data from clinical testing of the innovator product. The number of trials required for a biosimilar product varies by case, depending on the extent to which the manufacturer demonstrates that its biosimilar is comparable to the reference product.

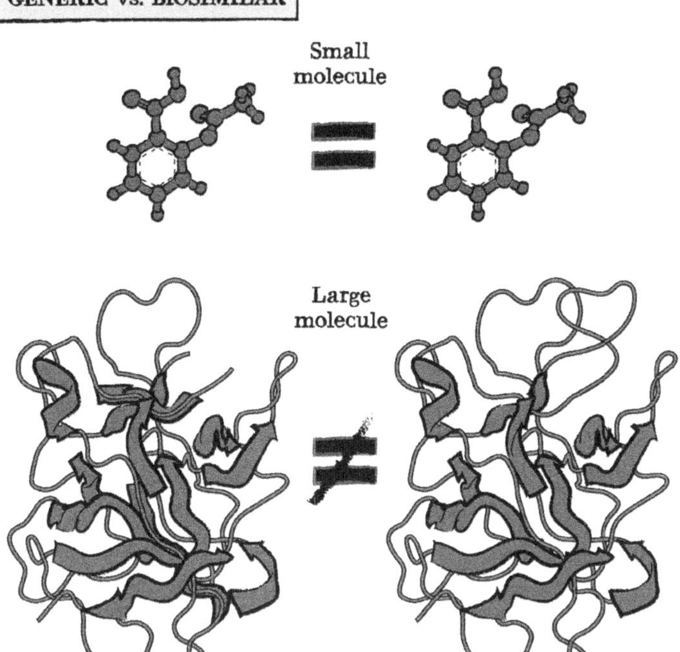

Tricky Terms

Drug Targets

No matter their size or type, all drugs are designed to affect just the right molecule involved in the disease pathway. Specifically, drug development scientists seek out exactly the infinitesimal part of our cells on which a disease process hinges. These molecules are called **drug targets;** mostly they're proteins. Drug targets are where the drugs attach to mitigate disease.

INDUSTRY NOTE: Data Exclusivity

Drug discovery companies typically seek patents early in the discovery process to lock out competitors. If the U.S. Patent Office grants a patent, the company receives twenty years of exclusive rights to make and to sell the product. However, compared to the development of other kinds of goods, the pharmaceutical development cycle is protracted. It consists of long years in early drug discovery and preclinical testing, followed by as much as a decade in human clinical testing. As a result, often only a few years of patent protection remain once a drug finally gains approval.

To compensate for this reality, the FDA grants "data exclusivity" to companies with newly-approved drugs. For a set time after approval, the FDA won't OK a generic or biosimilar version of the new drug based on the manufacturer's original safety and efficacy data. The data exclusivity period varies. Small molecule drugs receive five years, with an additional six months for pediatric drugs. Medicines for "orphan" diseases (illnesses that affect fewer than 200,000 people in the U.S.) receive seven years of exclusivity. Biologic drugs benefit from an even longer data exclusivity period—twelve years. This length of time reflects the markedly greater manufacturing costs of biologics and a correspondingly longer amount of time required to recover their development costs.

INDUSTRY NOTE: GAIN

In 2012, Congress passed legislation aimed at spurring the development of new antibiotics to fight infections that are resistant to existing drugs. The Generating Antibiotics Incentives Now (GAIN) Act defines certain new medicines as "Qualified Infectious Disease Products" or QIDPs. These antibacterial or antifungal drugs specifically treat serious or life-threatening conditions. QIDPs can receive fast-track and priority review status. They also earn an additional five years of data exclusivity. Thus far, the FDA has designated 147 drugs as QIDPs and approved 12.

Drug Targets

Some molecules make better marks than others. Receptors and enzymes serve as excellent targets. Why? Because they both play vital roles in transmitting chemical messages from a cell's surface to its nucleus. This chemical message activates a specific gene, which in turn kicks off the production of the corresponding protein. This cascade is called signal transduction.

All important cellular processes, such as cell division, hormone secretion, or immune cell activation, rely on signal transduction. Accordingly, mutations in receptors or enzymes are often associated with illness. Focusing on them can help manage or even cure a disease. Let's look at the example of growth factor signaling.

Growth Factor Signaling: A Major Drug Target

Growth factors tell our cells to multiply. Cells only respond to those that match their surface receptors. For instance, **epidermal growth factor** (**EGF**) stimulates skin cells to grow during wound repair. Cells covering the gut, lung, and breast also express receptors for EGF. Other cells have receptors for different growth hormones.

When a cell gets the initial "call" from the growth factor, it triggers the enzymatic activity of the portion of the growth factor receptor that's inside the cell. That, in turn, switches on **kinase** activity. That's the receptor's ability to transfer a phosphate group from one molecule to another, a process called **phosphorylation**.

This phosphate transfer causes the protein *receiving* it to adjust its shape just a bit. That shape change typically turns on the recipient protein's own kinase activity. This newly awakened protein kinase then kicks on yet another protein kinase, and so on down the line. The whole process is called a **signal transduction pathway** or simply a **signaling pathway.** The last element to be phosphorylated is usually a protein called a transcription factor. Once that happens, this final protein enters the cell's nucleus. There it binds to the DNA, which turns on the expression of a specific gene.

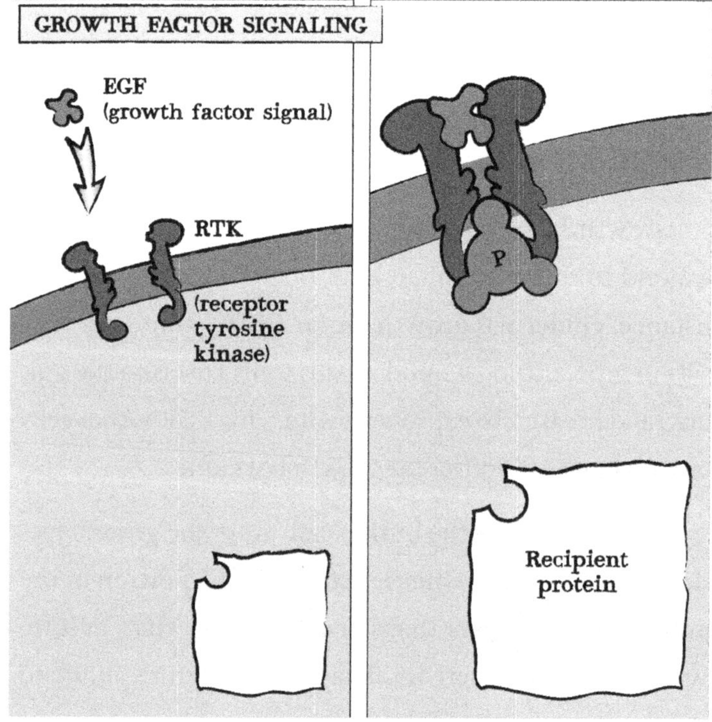

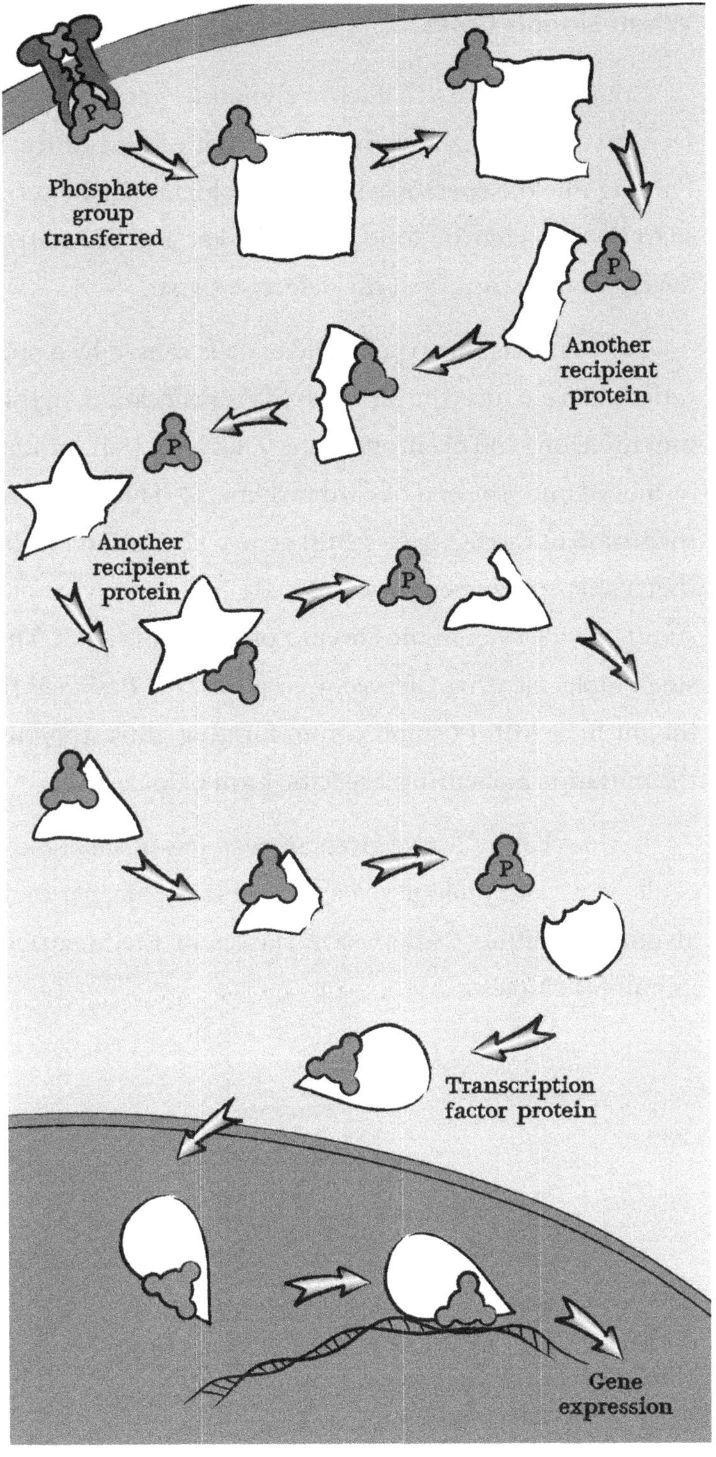
Phosphate group transferred
Another recipient protein
Another recipient protein
Transcription factor protein
Gene expression

When Signals Cross

Sometimes growth factor signaling gets bungled. Defective signals can be associated with various cancers. Finding out the specific signaling problem may help researchers to identify a new drug target and ultimately develop a drug to correct the defective signal.

For example, one type of leukemia is caused by a mutation referred to as the Philadelphia translocation. In this mutation, one end of chromosome 9 has been translocated or moved onto the end of chromosome 22. This results in the fusion of the BCR and ABL-1 genes. The fusion means that a kinase enzyme is always on—telling the cell to constantly divide even in the absence of a growth signal. The small molecule drug Gleevec was specifically designed to inhibit BCR-ABL-1 complex from forming, thus stopping the mutation and curing a specific form of leukemia.

In this chapter, we've given an overview of small molecule drugs and biologics. Next, we'll take a deeper look at one of the most innovative classes of therapeutics: immunotherapies.

CHAPTER 4

Immunotherapies: Harnessing The Power of the Immune System

Modulating Our Immune System

In the previous chapter, we introduced the two major classes of drugs: small molecule drugs and biologics. One kind of biologic—immunotherapies—work essentially by modulating our immune systems. Although one of the hottest areas in biopharma today, immunotherapies have a surprisingly long history.

In 1891, Dr. William Coley observed that some cancer patients went into spontaneous remission after developing skin infections. It turns out they'd been exposed to a strain of *Streptococcus A* bacteria. The doctor reasoned that dosing cancer patients with these microbes might trigger the same response. So was born "Coley's Toxins," a mixture of live and inactivated *Streptococcus* and *Serratia*. Injecting this bacterial blend successfully treated a range of cancers, including sarcoma, lymphoma, and testicular carcinoma. Scientists then knew far less about immunity than we do today, so they couldn't explain the treatment's success. The medical community perceived the toxins as too risky and they fell into disuse. For the next several decades, surgery and radiation took center stage in oncology.

Flash forward to the 1980s: immunologists have worked out the cellular basis of immunology. Further, they understand that our immune system can be a precise, powerful tool for healing. Scientists began to develop monoclonal antibodies (mAbs) as targeted therapeutics, first for cancer and then for other illnesses such as autoimmune disorders. In 1997, the FDA approved the first mAb, Rituxan, to treat non-Hodgkin's lymphoma. Drug companies continue to develop new mAbs, which remain some of the most potent therapies in biotech. More recently, they've been joined by powerful cell-based immunotherapies. Let's look at how they work.

Biotech's Bloodhounds: Monoclonal Antibodies

Monoclonal antibody (mAb) therapeutics zero in on an infinitesimal part of a protein associated with a disease, the **antigen.** In particular, mAbs target one of the antigen's structural sidekicks—an **epitope**. These are small clumps or strings of amino acids on which antibodies home in. But not just any epitope will do. Drugmakers design mAbs to find that certain special **epitope** and bind it.

This precision helps make mAbs extremely powerful. Because they glom onto only one target, they're much less likely to cause "off-target effects"—the adverse reactions that make patients miserable. As a result, they generally have very good safety profiles. mAbs have proven extremely effective in treating a wide range of diseases. The FDA has approved a number for various cancers and autoimmune diseases. They've also been ok'd for some infections and high cholesterol. As their versatility suggests, monoclonal antibodies aren't monolithic. They operate through diverse mechanisms, including provoking an immune response, blocking receptors, interrupting cellular signals, and even delivering toxic drugs. Here's how:

- mAbs trigger immune response: When a mAb latches onto an epitope, it acts as a flag signaling macrophages, cytotoxic T-cells, and other **leukocytes** to destroy the protein to which the epitope is anchored. In designing mAbs for epitopes/antigens on the outside of tumor cells, scientists have in turn created drugs that activate white blood cells to attack cancer. For example, the mAb Rituxan binds to the CD20 protein on the surface of non-Hodgkin's lymphoma cells, triggering the immune system to destroy those cells.

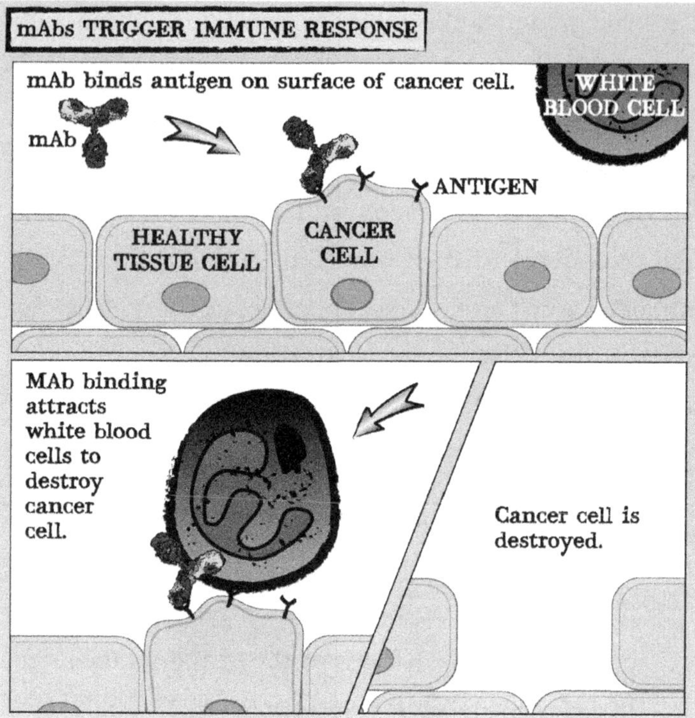

- mAbs block receptors: Recall that receptors are proteins on the surface of a cell. They send messages into the cell. For example, growth factor receptors transmit, you guessed it, growth signals. Overactive or overabundant growth factor receptors actually cause some cancers, such as HER2-positive breast cancer. Antibodies constructed to bind growth factor receptors "*without* turning them on" stop any growth signals. This prevents the cell from growing uncontrollably.
- mAbs capture signals: mAbs also block cellular communication by aiming for the signal itself. Creating antibodies to bind to signaling molecules still in the bloodstream prevents the signals from talking with

their receptors. For example, Humira, a mAb for autoimmune disorders, captures an inflammatory signaling molecule before it gets to its target receptor on the leukocyte surface.

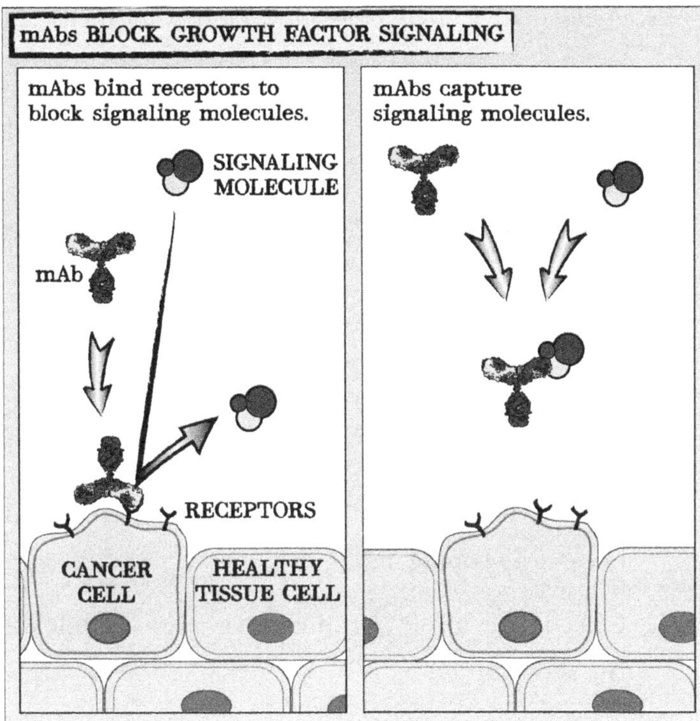

Special Delivery: Antibody-Drug Conjugates

The medical advances mAb technology delivers just keep coming. One of the latest permutations, **antibody-drug conjugates (ADCs),** melds the precision of monoclonal antibodies with cancer-killing toxic drugs. This one-two punch can destroy cancer cells while minimizing damage to healthy tissue.

ADCs have three main components:

- *A monoclonal antibody* that specifically targets a tumor-associated antigen that appears minimally on healthy cells or doesn't show up at all.
- *A highly toxic small molecule drug* to kill cancer cells once it gets inside them.
- *A linker molecule* that attaches the small molecule drug to the antibody.

How does an ADC work?

1. The antibody binds to its target antigen on the cancer cell surface.
2. The diseased cell "swallows" or internalizes the antibody-drug conjugate.
3. Once inside a cell, the linker degrades and releases the active drug.

The ability to target *only* malignant cells allows doctors to give people far more potent medicines than possible with traditional chemotherapy. The ADC's precision enables it to avoid harming healthy tissue.

Bispecific Antibodies

You may not realize it, but antibodies are Y-shaped. The "arms" of our natural antibodies and those of most therapeutic antibodies are structurally identical. They each recognize one target only.

However, genetic engineers have built a better antibody. By splicing genes from two different monoclonal antibodies, they've created a super-Y. These new, *bispecific antibodies* can detect two targets and bring them together. The first FDA-approved bispecific antibody, Blincyto, came on the market in 2017. Here's a summary of how it works:

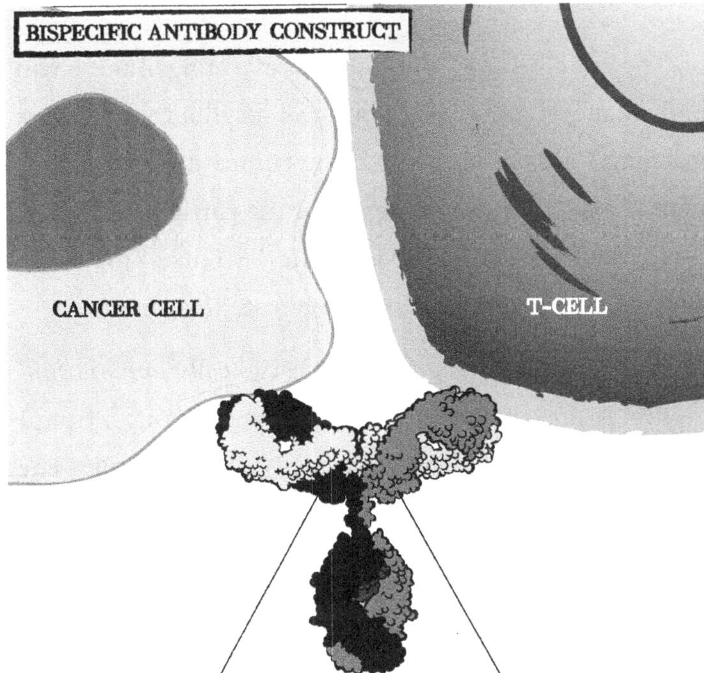

Imagine that one arm of the bispecific Y identifies and grabs a cancer cell. Meanwhile, the other recognizes and binds to a killer T-cell. (Remember, killer T-cells are white cells that inject toxins directly into cells.) By bringing the captive cancer cell into contact with a killer T-cell, the first arm of the Y enables the killer T-cell to annihilate the cancer cell.

Up to this point, we've primarily discussed how the medical community has tapped into immunotherapy's power against cancer. However, this approach plays a major role in other, less dire conditions as well. Let's look at how it helps control high cholesterol.

Checkpoint Inhibitors

The word "tumor" alone sounds alarming. But did you know that even the tumor's physical neighborhood causes problems? This area is called the **tumor microenvironment (TME)**. This phrase describes the particular physical locale in which a tumor exists. The TME itself may contribute to tumor growth and complicate treatment.

First and foremost, the TME is typically *immunosuppressive*. It prevents our bodies from recognizing a tumor as harmful, thanks to its own twisted cellular makeup. The TME contains "suppressor cells." These white blood cells, including tumor-associated macrophages and regulatory T-cells, emit signals that repress activated immune cells nearby.

BIOPHARMA INNOVATION: PCSK9 Inhibitors

When we're healthy and eating right, the body naturally keeps LDL cholesterol—the kind your doctor gets on your case about—in check with the help of low-density lipoprotein (LDL) receptors on the surface of the liver. Excess LDL bind to LDL receptors. Our livers absorb this excess LDL/LDL receptor complex, break down the cholesterol, and then sends the LDL receptor back to the surface of its liver cells. There, the recycled LDL receptors bind to more LDL, and the cycle repeats.

However, sometimes even a normal protein can cause issues. In this case, the troublemaker is excess PCSK9. This protein also binds to the LDL receptor. PCSK9, too, triggers the liver to absorb it and its LDL-receptor partner. Here's where the LDL recycling system can go awry. This time, the liver degrades *both* PCSK9 protein *and* LDL receptor—trashing the LDL receptors that would otherwise be recycled to the surface to continue to remove bad cholesterol. Fewer LDL receptors mean more dangerous LDL cholesterol flowing through our veins. This poses a problem, especially for those people who can't take statins.

Fortunately for them, researchers have come up with a mAb workaround with two new medicines. The PCSK9 inhibitors Praluent and Repatha attach to the PCSK9 protein and prevent it from mingling with LDL receptors on the surface of liver cells at all. This keeps the critical LDL receptors out of the cellular landfill, lowering LDL levels. This reduction in bad cholesterol lessens the risk of a heart attack or other cardiovascular problems.

Clinical results show that the inhibitors cut LDL levels by as much as sixty percent. These medicines lower cholesterol levels as effectively as statin drugs but with fewer side effects.

Tumors also release growth factors that nurture the growth of blood vessels feeding the tumor. Finally, the TME inhibits tumor-specific antigens on the tumor surface. These tumor-specific antigens are crucial for mobilizing killer T-cells. Disabling them further encourages the mass to grow.

Some tumors also activate immune system checkpoints. Checkpoints are proteins that shut down T-cells. Checkpoint proteins are activated by proteins on the surface of healthy cells. Turning on a checkpoint protein shuts down T-cells and helps our body defend itself against autoimmune disease.

PD-1 is a common checkpoint protein. It's activated by the protein PD-L1 on the surface of healthy cells. Unfortunately, some tumors display extra copies of PD-L1. Even if T-cells recognize the tumor, the tumor's PD-L1 proteins switch off the T-cells and prevent the T-cells from attacking the tumor.

Thankfully, the medical community has figured out a way to neutralize this menace. **Checkpoint inhibitors** are monoclonal antibodies that bind to the PD-1 protein on the surface of T-cells *or* to the PD-L1 protein on the tumor itself. This prevents PD-1 from interacting with PD-L1. If they can't get to each other, the T-cells remain active. Checkpoint inhibitor therapies enable the patient's own immune system to better fight malignant cells and start cleaning up the neighborhood.

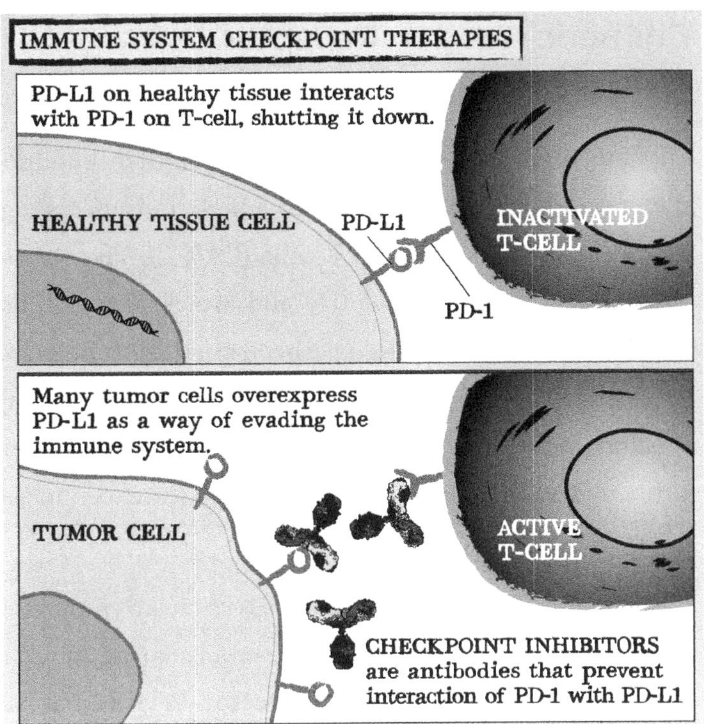
IMMUNE SYSTEM CHECKPOINT THERAPIES
PD-L1 on healthy tissue interacts with PD-1 on T-cell, shutting it down.
HEALTHY TISSUE CELL
PD-L1
INACTIVATED T-CELL
PD-1
Many tumor cells overexpress PD-L1 as a way of evading the immune system.
TUMOR CELL
ACTIVE T-CELL
CHECKPOINT INHIBITORS are antibodies that prevent interaction of PD-1 with PD-L1

Chimeric Antigen Receptor T-Cells (CAR-T)

Oncologists and patients also have another immunotherapy at their disposal. Even its name sounds menacing: **chimeric antigen receptor T-cells** (CAR-T). They boost the body's ability to recognize and attack cancer cells. These "super" killer T-cells have been physically enhanced to go after cancer. Like the mythical chimera, this drug is composed of different parts. Genetic engineers fuse an antibody with a killer T-cell receptor to create a chimeric molecule—the "C" in CAR-T.

The transformation begins with technicians removing killer T-cells from a patient's body and isolating them in the lab. Next, scientists use a **viral vector**—a virus that has been modified to contain a therapeutic gene. In this case, the viral vector delivers a gene that encodes the therapeutic antibody to the T-cells.

The enhanced receptor (T-cell/therapeutic antibody complex) includes two parts: a *targeting domain* and an *activation domain*. The first is the portion that remains on the surface of the T-cell. It's an antibody that detects and locks onto a specific surface protein on the patient's cancer cells. The *activation domain* part of the receptor is triggered once the targeting domain attaches itself to the desired cancer protein.

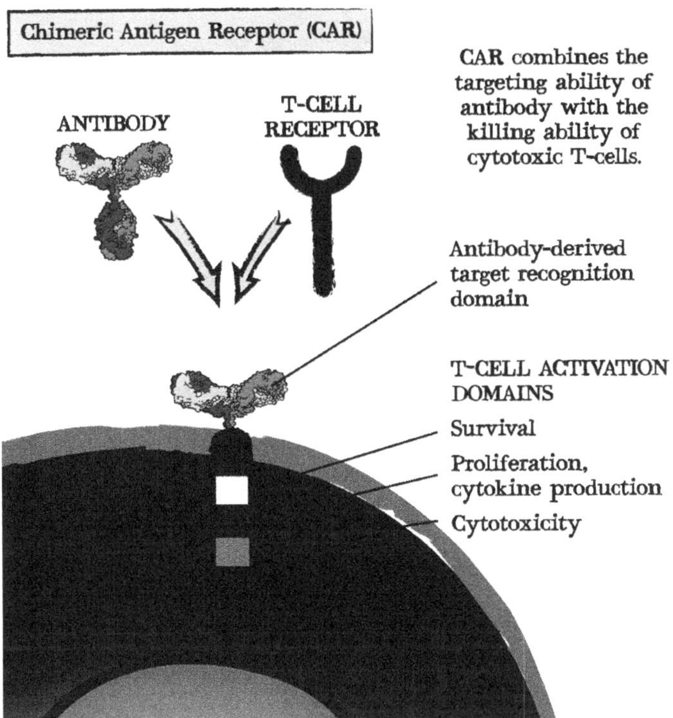

The engineered T-cells are then reinfused into the patient. Once back inside the patient, the targeting domain finds the proper surface protein on the tumor cell and attaches to it. Then, the activation domain signals the killer T-cell to:

- Stay alive.
- Make copies of itself (proliferate).
- Release **cytokines**—chemical signals that activate other white blood cells to assault the tumor.
- Kill the target cell.

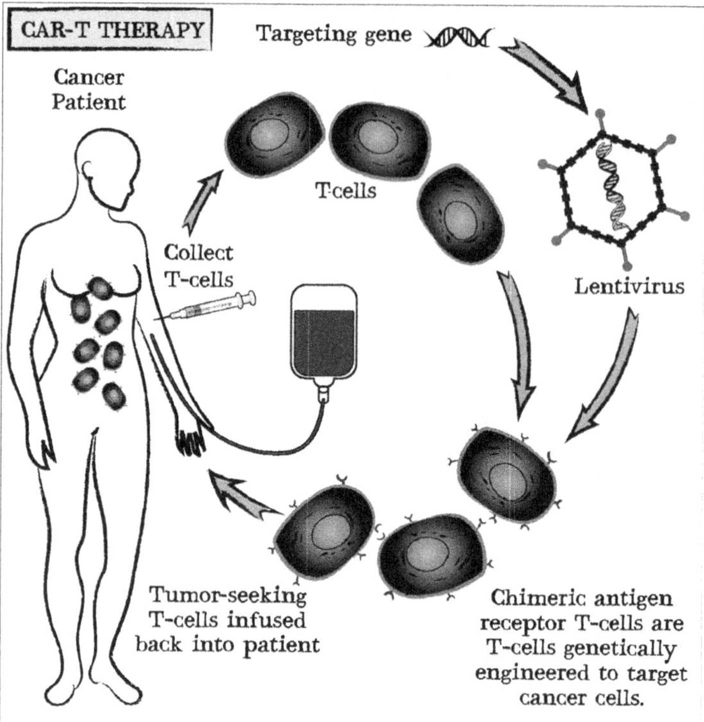
CAR-T THERAPY
Targeting gene
Cancer Patient
T-cells
Lentivirus
Collect T-cells
Tumor-seeking T-cells infused back into patient
Chimeric antigen receptor T-cells are T-cells genetically engineered to target cancer cells.

If you think CAR-T treatments sound a bit like search and destroy, you'd be right. The FDA has approved CAR-T therapy for certain blood cancers. It's successfully cured leukemia and lymphomas in patients whose disease didn't respond to other treatments.

Tricky Terms

CAR-T

The medical community classifies CAR-Ts as a "cell-based gene therapy." They're immune cells that have been engineered using gene therapy techniques. Kymriah, a CAR-T therapy approved in 2017, was the first gene therapy to be approved by the FDA.

Cytokine Storms

As potent as they are, CAR-T therapies carry serious risks. Using them raises the possibility of cytokine release syndrome (CRS) or a cytokine storm. **Cytokines** are small proteins that play an important role in relaying messages from a cell to surrounding cells and tissue. Cytokines serve two core functions:

- They activate additional white cells to fight pathogens.
- They stimulate white blood cells to move towards inflammation.

Cytokine signaling makes for a quick, strong, and usually appropriate immune response via a positive feedback loop in which activated white cells release more activating cytokines. The response typically dissipates when the pathogenic cells have been destroyed. However, sometimes the feedback loop just keeps looping. This phenomenon is called a cytokine storm.

CRS consists of acute inflammation with high fever, swelling, and/or nausea. It sometimes causes serious tissue damage and even death. For example, excess fluids can seep into the lungs and cause them to fail. Although uncommon, cytokine storms pose the biggest risk of CAR-T treatments.

Calming the Storm

Biopharma companies are working on new CAR-T treatments with controls to regulate cytokines and minimize storms. First-generation CAR-T therapies induce maximum white cell activity—a full cytokine barrage. Their intensity makes tamping down the cytokine response impossible.

In one example of the next-generation therapy, patients take an adjunct small molecule drug during their CAR-T infusion. This sidekick medicine provides an on/off switch for CAR-T therapy. It turns on the treatment to fight cancer. Should a cytokine storm ensue, doctors can immediately withdraw the adjunct drug, turning off CAR-T and ultimately stopping cytokine release.

The idea of bifurcated bispecific CAR-T cells presents another possibility for preventing cytokine storms. The concept arises from the difficulty of finding target proteins that appear *only* on cancer cells. The technique equips each engineered killer T-cell with not one but two chimeric antigen receptors (CARs). Here's how it works:

- One CAR activates in the presence of a certain protein on the surface of cancer cells. The CAR-T cell then produces more copies of itself, releases cytokines, and attacks the tumor.

- A second inhibitory receptor (iCAR) activates in the presence of a protein found only on healthy cells. This means that if the CAR-T is wrongly triggered by the presence of a tumor antigen on the surface of a healthy cell, the inhibitory receptor is also be triggered, shutting down CAR-T activation.

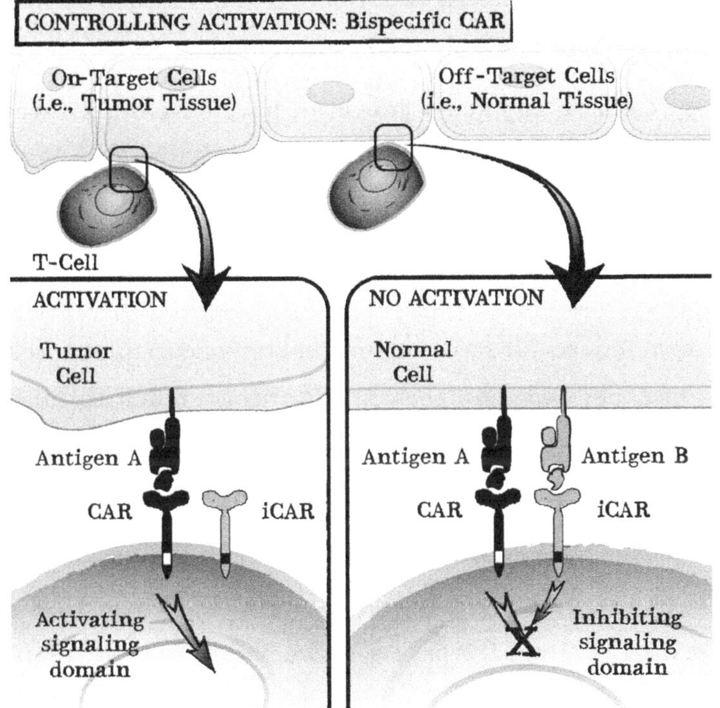

CAR-NK and CAR-MA

Our bodies also rely on the built-in protections of our innate immune response, such as natural killer (NK) cells. Many immunologists believe that prompt action by NK cells helps eliminate many cancer cells before they ever become problematic. Sometimes, however, early-stage tumors display so few abnormal proteins on their surface that the immune system fails to respond.

Genetic engineers, however, have developed a way around this hole in our oncological defenses—one with significant advantages over CAR-T treatments. Researchers at the University of Texas MD Anderson Cancer Center have constructed NK cells to display a CAR. This enhancement induces NKs to recognize and respond to tumor cells. Once they switch on, the CAR-NKs behave much like killer Ts, releasing cytokines that bolster the immune response—killing the nasty cells by injecting even nastier toxins.

CAR-NK cells possess three important advantages over CAR-T cells: safety, accessibility, and effectiveness. CAR-T cells come from the patient's own T-cells to evade graft-versus-host-disease (GVHD). This potentially fatal illness strikes when a patient's immune system rejects foreign tissue. Donor *NK* cells, in contrast to donor *T*-cells, don't appear to cause GVHD.

Medical professionals can obtain NK cells relatively easily—for example, from donated blood or from umbilical cord blood. Labs modify donated cells to express a CAR, which is then injected into the patient. In contrast, removing and engineering a sick patient's own T-cells, then reinfusing them costs much more money and time. The greater availability of donor cells may make it possible for companies to create "off-the-shelf" products for CAR therapy more readily than for CAR-T therapy. In addition, lower costs may mean more treatments reach those who need them. Finally, preclinical research at the University of California, San Diego suggests that CAR-NK may attack solid tumors more effectively than CAR-T, which most successfully treats leukemia and other blood cancers. CAR-NK treatments are beginning to enter clinical studies.

Chowing Down On Cancer

Another more visceral approach to immunotherapy involves hijacking our bodies' scavengers—macrophages. The name derives from a Greek phrase meaning "big eater." These cells kill infectious or diseased cells by surrounding and digesting them. Think of a sloppy Pacman. After a macrophage meal, minute bits of foreign cell proteins or antigens remain on its surface. The leftovers help activate some of the immune system's other defenses, such as killer T-cells.

Researchers are exploring the possibility of constructing CAR-macrophages to destroy specific cancer cells. The amped-up macrophages will simultaneously trigger other immune cells to recognize and attack those same antigen-bearing cells. Like other macrophages, CAR-macrophages can penetrate solid tumors much more effectively than basic T-cells. Perhaps one day, they may work even more powerfully against cancer than CAR-Ts.

Tricky Terms

Oncolytic Virus

An **oncolytic virus** has been engineered to selectively infect and kill cancer cells. This beneficial virus takes out cancer in two ways. The first is through lysis, the explosion (well, bursting) of cells that results from viral multiplication. The second comes through an immune response targeting the tumor cells prompted by the infection.

Turning Down Immune Response: CAR-Treg

Scientists, physicians, and others mostly discuss CAR therapies in the context of bolstering the immune system to fight cancer by engineering cytotoxic T-cells, NK cells, or macrophages. But what about turning *down* immune response? Researchers have begun to think about using CAR technology to treat autoimmune and inflammatory diseases with regulatory T cells—Tregs—that target affected tissues. A Treg is a type of T-cell that releases anti-inflammatory cytokines when activated. Consider, genetic engineers could create Tregs that target intestinal cells of Crohn's disease patients. The Tregs could also be administered to a patient's gut and release anti-inflammatory cytokines, relieving the symptoms of irritable bowel disease.

This chapter discussed how immunology translates into therapies designed to fight cancer and inflammatory disorders. As our understanding of the immune system and its interplay with disease continues to expand, so too will these therapies. In the next chapter, we discuss the original immunotherapy: vaccines.

CHAPTER 5

Vaccines: Elegant, Powerful Simplicity

Beginning in 2020 hundreds of thousands of people worldwide came down with COVID-19. To date 3.4 million people have died (World Health Organization) from this coronavirus. Every year in the United States, tens of thousands of people die from influenza. The 2017-18 season was especially rough. Nearly 49 million people fell ill. Almost 80,000 of them died according to the U.S. Center for Disease Control and Prevention (CDC). Other than wearing masks and washing our hands, what's the most effective COVID-19 and flu preventative? The humble vaccine. It's modern medicine's original immunotherapy.

This chapter examines the basics of vaccination, including its development and manufacture. We then look at some of the latest innovations in vaccine technology.

Cocktail Fodder

40,000 Saved!
According to the CDC, flu season runs from October through May in the U.S. The organization's latest study estimates that the flu vaccine has saved 40,000 lives over nine years.

Jenner's Needle in the Haystack

Vaccines have been around a while—dating from the late 18th century and the smallpox pandemic. This disfiguring disease killed or disabled hundreds of thousands of people in England alone. In1796, English doctor Edward Jenner noticed that local milkmaids were immune to smallpox. Jenner observed that these women had all had cowpox, a related but harmless virus. He hypothesized that cowpox imparted some type of immunity. To test his theory, Jenner took pus from a cowpox blister and inoculated his gardener's small son, James, with the virus through shallow scratches. James developed a slight fever. Later, when intentionally exposed to smallpox, the little boy never fell ill.

Jenner's methods were a little rough (not to mention unethical by today's standards), but his thinking was spot on. The idea behind vaccines is simple. First expose someone to a harmless version of a disease-causing microorganism, or pathogen. Amazingly, this "trains" their immune system to recognize and fight the germ. The exposure forces the body to create special white blood cells, known as memory cells. If the person encounters the pathogen again, their bodies fights the infection much more robustly than if there had been no previous exposure.

Types of Vaccines

Vaccines come in different varieties including inactivated whole, live attenuated, and subunit vaccines. Each requires different manufacturing requirements.

- Inactivated whole vaccine. Made with dead microorganisms (viruses or bacteria), these stimulate an immune response. Among the most famous is Dr. Jonas Salk's polio vaccine, developed in 1955.
- Live attenuated (weakened) vaccine. These are created by reducing a pathogen's strength so that they become harmless. Live vaccines tend to produce the strongest immune reaction.
- Subunit vaccine. These vaccines use only one part of a pathogen; a protein called an antigen. The antigen provokes an immune response. One production method involves isolating an antigen from a virus and administering only this protein.

Scientists are also currently developing DNA-based vaccines. These consist of a gene encoding a pathogenic protein as opposed to the protein itself.

The Whole Story

Most whole pathogen vaccines protect against viruses such as measles or mumps, not dangerous bacteria. Making any vaccine means first growing lots of viruses. "Virus-farming" involves selecting a strain of a particular virus, aka the seed strain, and choosing what to grow it in (the medium).

Where do manufacturers buy a pathogen in the first place? They come from one of two sources. Viruses (and other microorganisms and biological materials) are produced and housed in well-established culture collections, such as the American Type Culture Collection (Manassas, VA.) Some companies or academic institutions also develop "in-house" strains of particular viruses. For the flu vaccine, new strains are selected annually based on the World Health Organization's assessment of how the virus has changed over the previous season.

Choosing the seed strain is the first step of vaccine manufacture. Although sometimes dangerous, even deadly, viruses are powerless without a host—someone or something to grow on. Two of the most common "host-cell platforms" are chicken eggs and animal cell culture.

Eggs? Yes, the incredible edible egg provides a fantastic growth medium for influenza and other viruses. (Side note: The CDC recommends that most people with egg allergies be vaccinated. However, it also suggests that anyone who's experienced a severe allergic reaction to the flu vaccine does *not* get one.)

Some viruses thrive in certain types of animal cells. Two of the most commonly used in vaccine production include one from the kidney of the African green monkey—known as Vero cells. The other, MDCK cells, originate from the kidney cells of a cocker spaniel. Though it's easier and quicker to scale up animal cell culture production, it's much pricier than egg-based vaccine growth.

Once the manufacturer has enough virus, it must be separated or isolated from the host material. Isolation involves centrifuging and filtering to divide virus particles from host cells.

The next step in production of whole pathogen or inactivated vaccines—inactivation—is crucial. Manufacturers disable a virus's ability to infect without eliminating the portion that triggers an immune response. Inactivation occurs through a variety of strategies, including:

- Detergent-treatment. Used exclusively on enveloped viruses (those with an outer lipid membrane), detergents break the chemical bonds that hold the viral envelope (outside surface) together, making it unable to invade a host.
- Heat, chemical, and pH treatments. Viruses use proteins on their surface to infect their hosts. Altering their shape destroys their ability to recognize and get into cells.
- Ultraviolet light-treatment. UV light destroys a virus's building blocks—their DNA or RNA. Without genetic instructions, viruses can't replicate.

The last step in making vaccines is formulation. The inactivated virus gets mixed into a sterile water or salt solution with stabilizers and preservatives. Manufacturers bolster some vaccines with **adjuvants** at this point. Adjuvants are ingredients such as aluminum salts that boost the immune response to a vaccine. Vials are then filled, inspected, labeled, and shipped. Most vaccines require refrigerated storage and distribution.

Alive, Not Kicking

Making a live attenuated vaccine follows similar steps but *without* inactivation. The process starts with a seed strain that has been rendered harmless. The new organism continues to grow but produces immunity without causing illness. Attenuated vaccines produce stronger, usually longer-lasting, immune responses than inactivated vaccines because they more closely mimic actual infection. Doctors recommend that older people, cancer patients, and others with weakened immune systems not get attenuated vaccines.

Companies manufacture attenuated vaccine strains in a few different ways. Sometimes, a vaccine may consist of a related but harmless virus that kicks in the immune response. Jenner's vaccine, which was essentially cowpox, provides the classic example. Today's smallpox vaccine uses a related virus, *vaccinia*.

Another common method involves raising several generations of a *clinical isolate*, or a laboratory-pure, pathogen. Growing in non-human cells, it adapts to its new host, becoming less infectious over time. Examples include measles, yellow fever, and poliovirus vaccine strains. Scientists can also use recombinant DNA technology to delete the portions of a viral genome that cause infection.

Vaccines rest on a simple premise, but the science and manufacturing that make them possible are complex. Different microorganisms require individual approaches. Through trial and error, microbiologists, virologists, and other scientists determine the best formulation.

When the Part is Greater Than the Whole

The structure of a virus is much simpler than that of our cells. Despite the damage they can inflict, viruses consist only of one or a few strands of DNA or RNA encased in a protein shell.

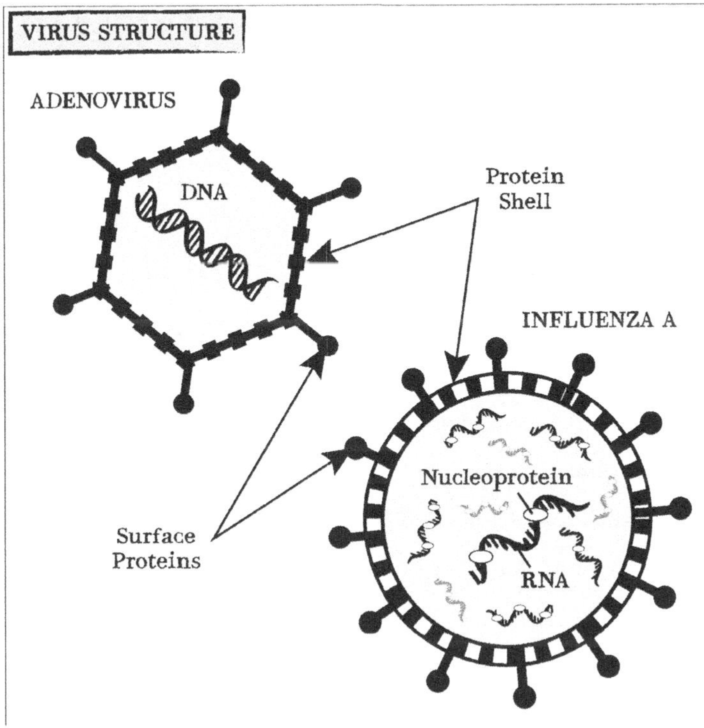

This relatively straightforward structure means that sometimes only part, or a subunit, of a virus suffices to stave off infection. Different subunit vaccines use different bits of a pathogen. Often, they consist of nothing more than one viral surface protein. Subunit vaccines work because our immune systems recognize and respond readily to these surface proteins.

For these vaccines, drug manufacturers alter yeast, bacteria, or Chinese hamster ovary (CHO) cells to produce a protein. They do so by equipping them with the gene encoding the viral protein. These host cells then make the viral protein, which the manufacturer isolates and formulates into the vaccine. Subunit vaccines include those for hepatitis B and human papillomavirus (HPV).

Sometimes a viral subunit forms what is called a **virus-like particle (VLP)**–a protein structure with a shape that closely mimics a virus but without its genetic material. In these cases, the body's immune system responds very robustly.

A subunit vaccine can also derive from toxins produced by dangerous bacteria. For example, *Clostridium tetani* (tetanus) secretes tetanospasmin, a neurotoxin that causes severe muscle spasms, potentially leading to death. The vaccine contains inactivated toxin, which helps develop antibodies to prevent future illness. Because subunit vaccines contain none of a pathogen's genetic code, they're generally very safe.

Carbs + Protein=Immunity (Sometimes)

In general, it's harder to come up with vaccines that protect against bacterial infections than for viral infections. That's because the surface of some bacteria is covered in long chains of carbohydrate molecules. Called polysaccharides, they mask the bacteria's proteins. This cloaking device prevents the body from recognizing the pathogen's threat and mounting an immune response. But molecular biologists and other scientists have discovered that it's possible to link polysaccharides to a harmless protein, and thus coax an immune response. These vaccines are known as conjugated polysaccharide vaccines because the carbohydrate is conjugated (connected) to a protein. The *Haemophilus influenzae* type B (or Hib) and some pneumonia and meningitis vaccines are made this way.

Getting Down to the Essentials

The basics of immunization have been around over a century—use a disease-causing microbe, or just a part, against itself. However, the latest step in the evolution of vaccines takes a different tack, delving more deeply into the building blocks of life—DNA.

Instead of a whole pathogen or fragment, a DNA vaccine contains just a small bit of a virus's genetic code. Drug companies nestle the code in **plasmids**—small, circular DNA molecules commonly used in biotech to transfer genes into cells. As you can see below, the introduced DNA prompts the host to produce the target viral protein and consequent immune response within their own cells, but without an infection.

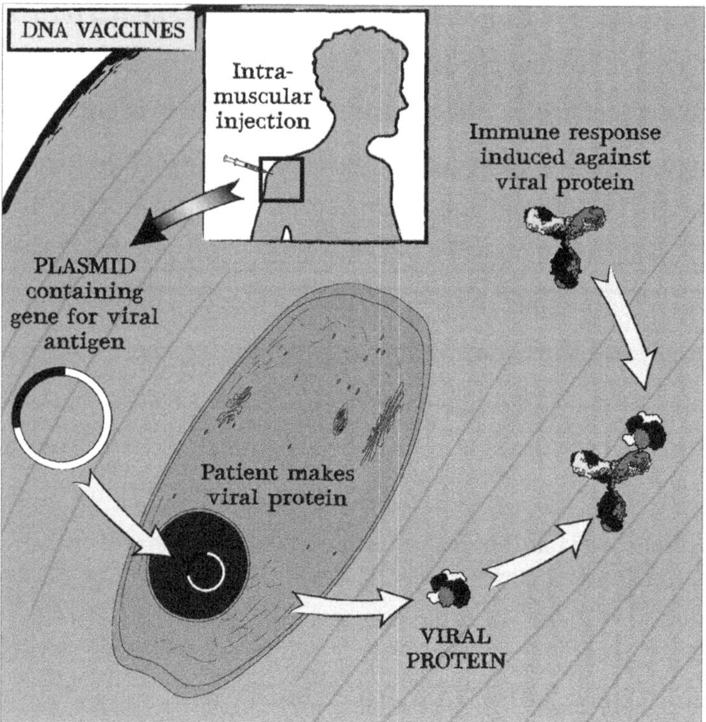

The big challenge for DNA vaccines is getting patients' cells to accept the new DNA. The most effective technique so far seems to be **electroporation**–administering the vaccine along with short pulses of electrical current. The electricity creates temporary pores in a patient's cell membranes, enabling the DNA to enter.

As of Winter 2021, the FDA had yet to approve any DNA vaccines for human use. Nevertheless, several are in development for hepatitis B and C, and the Ebola, HIV, and Zika viruses. The prospect of DNA vaccines offers important advantages, which include prompting a strong immune response and somewhat easier manufacturing. Producing large volumes of viral gene-containing plasmids still means growing lots of bacteria to reproduce the plasmids. However, purifying and formulating DNA vaccines is more straightforward due to the relative simplicity of the molecules' structure. These vaccines require no refrigeration, which extends their shelf life and the time available to transport them.

mRNA Vaccines and Covid-19

mRNA vaccines are another genetic approach to vaccines. mRNA vaccines have been in development for the past 20 years. Fortunately, mRNA vaccine technology has matured and improved to the point that multiple companies, including Moderna and Pfizer/BioNTech, could produce effective mRNA vaccines in 2020 to help combat the Covid-19 pandemic. This vaccine platform relies on messenger RNA (mRNA). mRNA is the molecule that cells naturally use to copy genetic information from DNA and translate into a protein. When a patient receives an mRNA vaccine, their cells translate the mRNA into new viral proteins. The patient's white blood cells then learn to recognize that protein, resulting in immunity to the virus. Encouraging cells to produce a viral protein very closely mimics natural infection. This pathogenic provocation can produce a strong immune response that engages both B- and T-cells.

The mRNA approach to Covid-19 is based on the mRNA sequence for the spike protein. Spike protein is a large, protruding glycoprotein that naturally sits on the outer envelope of SARS-CoV-2. As soon as immunologists observed this large, protruding spike protein, they knew it would likely make an excellent antigen, capable of provoking a robust immune response. In general, large, protruding molecules on the outside of pathogens tend to successfully activate our immune cells. Another key innovation, which has proved crucial for the mRNA vaccine, is the use of lipid nanoparticles (LNPs). Both Moderna and Pfizer/BioNTech rely on LNPs to serve as delivery vehicles for their mRNA vaccines. LNPs consist of a tiny ball of specialized lipids (i.e. fat) molecules. mRNA for the viral spike protein is enveloped within these lipid nanoparticles. When we receive an injection into the muscle tissue of our arm, the LNPs safely and effectively deliver the mRNA inside the membrane of our muscle and immune cells. Once inside our cells, the mRNA molecule can translate into viral spike protein. The viral spike protein then engages with multiple immune cells to create long term immunity.

The Future of mRNA Vaccines?

Because mRNA molecules are relatively straightforward to synthesize, this type of vaccine could incorporate mRNA encoding proteins from several different flu strains. The presence of many proteins would make the vaccine more universally protective. A number of different companies have mRNA flu vaccines in clinical and preclinical development.

Cocktail Fodder

Vaccination Activation

Getting a shot won't make you sick. Sometimes, people feel mild symptoms such as fatigue, headache, or a low-grade fever afterwards. The discomfort signals that the vaccine has triggered an immune response.

Testing, Testing

How does a new vaccine come to market? The FDA requires companies to test vaccines for safety and effectiveness on human subjects. The process differs somewhat from clinical drug trials. Researchers test a new drug on sick volunteers to see if it cures them; they administer an investigational vaccine to healthy volunteers to see if it prevents illness.

Clinical vaccine trials involve three steps:

- **Phase I** is typically a small study in which healthy volunteers get the vaccine. Doctors monitor them for side effects. If no unacceptable reactions occur, the vaccine advances to the next stage.
- **In Phase II** more subjects, typically hundreds, get the vaccine. The same number of people serve as a control group, which doesn't receive the vaccine. Both groups share the same level of risk for contracting the target disease. Researchers observe the

"vaccinatees" for two years or more to see if they contract the disease at lower rates than people in the control group. Researchers also monitor immune response by measuring anti-pathogen antibodies in the participants' blood.

- **Phase III** studies involve even more subjects—often thousands. These are people at high risk for the disease. They receive the vaccine and are monitored as described in Phase II, for from three to five years. This trial also includes a control group.

An Alternate Flu-niverse

In 2018, the National Institute of Allergy and Infectious Disease (NIAID) announced a strategic plan to create a vaccine that better protects against multiple strains of the influenza virus for multiple years. In other words, a universal flu vaccine. The Gates Foundation (Seattle, WA) has earmarked up to $12 million to support the project. Let's check out some of the science behind these initiatives.

The Epitope Hope

Flu prevention today relies on whole-virus vaccines. They deliver an inactivated version of the virus, which elicits an immune response in the person rubbing her or his sore arm. Vaccines typically create antibodies against the outermost portion of the hemagglutinin (HA) protein, which is found on the surface of the virus and helps it infect target cells. Our immune system recognizes the "head" of the protein most readily. Regrettably, that's the spot that mutates most rapidly—meaning that an immune response against it typically lasts for only one flu season.

One strategy for creating a universal vaccine targets the HA protein's "stalk." This bit of the protein mutates much less frequently than the head. Slower mutation means the stalk is much more likely than the head to remain the same from year to year. One of the most promising investigational efforts has yielded a **peptide vaccine**. It consists of short stretches of the HA stalk protein called "**epitopes**," or sequences known to induce an immune response.

Tricky Terms

Pandemic Strains

The World Health Organization defines a pandemic as The worldwide spread of a new disease. Devastating pandemics have emerged throughout human history. An 1817 Cholera outbreak killed millions. The Spanish flu of 1918, caused by an H1N1 virus with genes of avian origin, killed an estimated 50 million people worldwide.

Pandemic strains of the flu virus typically result from a pathogen that historically infects animals such as pigs or birds but then mutates to infect people. We use the term zoonotic disease to describe a disease created by a virus that transfers from animals into humans. Since these viral strains are brand new to the human population, few if any people have immunity. Thus, the illness spreads widely and rapidly. Protection against these potentially devastating flu strains tops the agenda of any universal vaccine researcher.

The Covid-19 outbreak in Wuhan, China was initially called an epidemic. Epidemics are a sudden increase in cases in a localized, geographic area. As the SARS-CoV-2 virus spread across several countries and affected a large number of people, Covid-19 then became classified as a pandemic.

Tricky Terms

Neoantigen

An antigen is a protein or portion of a protein in a cell that the immune system recognizes. Think of antigens as flags; some signify "all is well" and some spell trouble. An immune response occurs when white blood cells, such as macrophages and cytotoxic T-cells encounter a flag that means trouble. The best-case scenario involves immune cells recognizing, targeting, and killing cells with worrisome antigens. Worst-case scenario: cells with suspicious antigens remain undetected because they've evolved to operate in stealth mode. Cancer is often characterized by such a cloaking mechanism—which explains why it can be a silent disease until its very late stages.

As a tumor grows, it often accumulates additional mutations. Scientists have found some are significant enough to produce new antigens which the immune system *can* recognize. These **neoantigens** are the secret sauce in cancer vaccines.

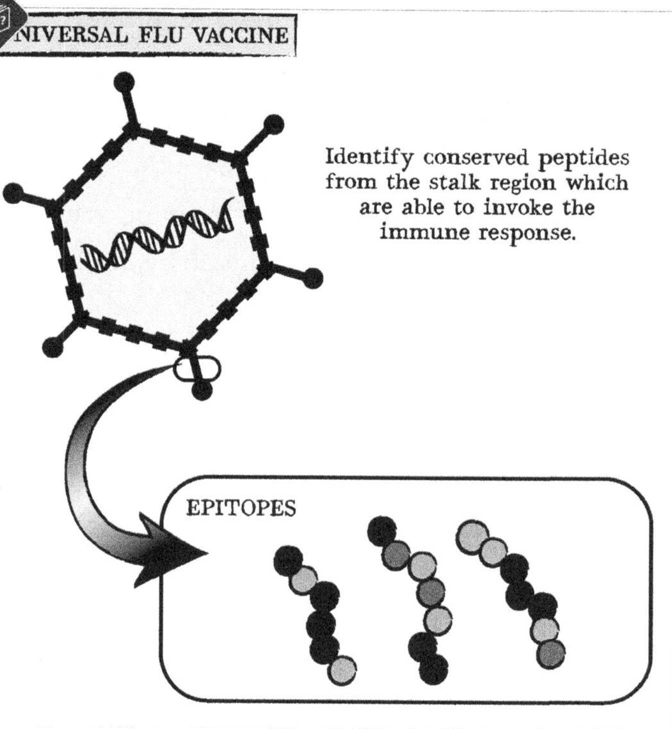

The nine epitopes selected for the vaccine are *conserved*, which means they're present on different HA proteins across different strains. So if this vaccine successfully prompts an immune response, it should protect against different strains of the flu over multiple seasons.

Preliminary data suggests that this blend of epitopes will induce not only a vaccine's usual antibody response but a T-cell response as well. The current flu shot doesn't. Once activated, T-cells quickly kill infected cells.

Can mRNA Save the Day?

The Elusive Cancer Vaccine

We're very familiar with the vaccines that our children cry over or that many of us got as kids: chicken pox, MMR, meningitis, and the rest. But what about a vaccine for the Emperor of All Maladies? The unfulfilled hope of cancer vaccines persists. Yet biopharmaceutical companies remain undeterred in pursuit of this prize. A new crop of biotech firms is utilizing the latest advances in genomics as they push to come up with vaccines that train the immune system to recognize and fight established malignancies. The key lies in identifying which aspects of tumors differentiate them enough from healthy tissue for the immune system to see them as foreign.

Finding Neoantigens

One of the hallmarks of cancer cell development is a high rate of DNA mutation. An established tumor may grow to display dozens or even hundreds of mutations that set it apart from healthy cells. Identifying mutations begin with performing a biopsy to collect and study a small sample of tumor cells. Thanks to advances in genome sequencing technology, pathologists can sequence the tumor's entire "exome"—the portion of the DNA used to make proteins. Accordingly, scientists need only look here for any neoantigens. They then compare tumor cell exomes to healthy cell exomes, to identify differences in DNA sequences.

Unsurprisingly, not all proteins make good antigens. The best are those displayed on the cell surface where the immune system can spot them. To flag cell surface neoantigens, researchers enter the "unique to cancer DNA sequences" into computer programs that predict the probability of surface location. Typically, around five percent of the mutated genes code for potential neoantigens.

Training the Vaccine

Once neoantigens are identified, they are synthesized in the lab and mixed with an immunity-boosting adjuvant. Ideally, a neoantigen-based vaccine contains at least twenty different neoantigens. This is to both produce a strong immune response and reduce the likelihood of resistance. A tumor may mutate and stop producing one neoantigen, but it's unlikely to stop producing several simultaneously.

Although in some cases, several different people may share common neoantigens, others may be unique to a single person. Neoantigen-based vaccines may accordingly end up being designed for one patient alone. This precision was unimaginable even just a few years ago. It's within reach today because of the increased efficiencies in both time and money achieved with genome sequencing.

Game Changers

Several neoantigen-based vaccines are in clinical testing. Currently, it takes around six to twelve weeks to identify neoantigens and produce a vaccine. The goal of oncologists working on these vaccines is to reduce this time to one month. Neoantigen-based vaccines in clinical development target breast cancer, melanoma, glioblastoma, and non-small cell lung cancer. If any succeed, we'll be witnessing truly personalized medicine.

In this chapter, we've reviewed the basic science behind vaccination, the different vaccine types, and some of the latest vaccine innovations. In Chapter 6, we leap forward into some of medicine's most cutting-edge treatments: gene therapy.

CHAPTER 6

Gene Therapy: Delivering on the Promise

The field of **gene therapy** first arose in the 1980s, as the study of molecular genetics emerged from university laboratories to start a burgeoning industry. It began with a revolutionary shift in how the scientific community understood hereditary disease: namely, the fact that a mutation in a single gene causes some diseases. Examples include sickle cell anemia, cystic fibrosis, and hemophilia.

The Miracle Preschooler

What if you could somehow change a flawed gene so that it works and cure the disease? That's the premise of gene therapy. The new gene itself becomes a drug. The first "proof of principle" for this concept came in 1990, when four-year-old Ashanti DeSilva was given gene therapy at the National Institutes of Health (NIH) for severe combined immunodeficiency disorder (SCID).

This condition completely disables a person's immune system. It arises from a defect in a gene required to produce white blood cells and results in life-threatening complications to minor infections. After Ashanti received a working copy of the gene, her immune system began to function normally. Ashanti is now in her early thirties and leads a healthy life. In the months following the first procedure, another young girl was also successfully given gene therapy for SCID.

Tragic Setback

Sadly, these early successes were followed by a major setback in 1999. Jesse Gelsinger, a teenage wrestler from Arizona, suffered from ornithine transcarbamylase deficiency. This heritable disease stems from a mutation in a gene required to break down ammonia, leading to loss of coordination, lethargy, and sometimes death. During a gene therapy clinical trial at the University of Pennsylvania, Jesse's immune system responded violently to the **viral vector** used to move the new gene into his liver cells. He died within days. Jesse's death caused the NIH to impose a moratorium on gene therapy trials and establish new monitoring plans to strengthen protection for trial participants. It also called attention to the critical importance of vectors—the microscopic vehicles that transport repaired genes into a patient's body—in gene therapy safety. In the years since, vector technologies have advanced considerably leading to new breakthroughs in gene therapy. Let's take a closer look at gene therapy vectors.

Special Delivery: Viral Vectors

From influenza to Ebola, we normally think of viruses as hazardous. In the case of gene therapy, however, some of these microbes benefit us. Molecular biologists have adapted viruses to deliver therapies by tweaking them to pursue disease instead of causing illness. These souped-up microbes are known as **viral vectors**. A virus is simply a segment of genetic material—RNA or DNA—surrounded by a protein coat. Proteins on the surface of the vector target proteins on the surface of a specific diseased or malfunctioning cell. The viral vector enters the target cell carrying the therapeutic gene. Ultimately, this enables a patient to make the missing functional protein, curing the disease.

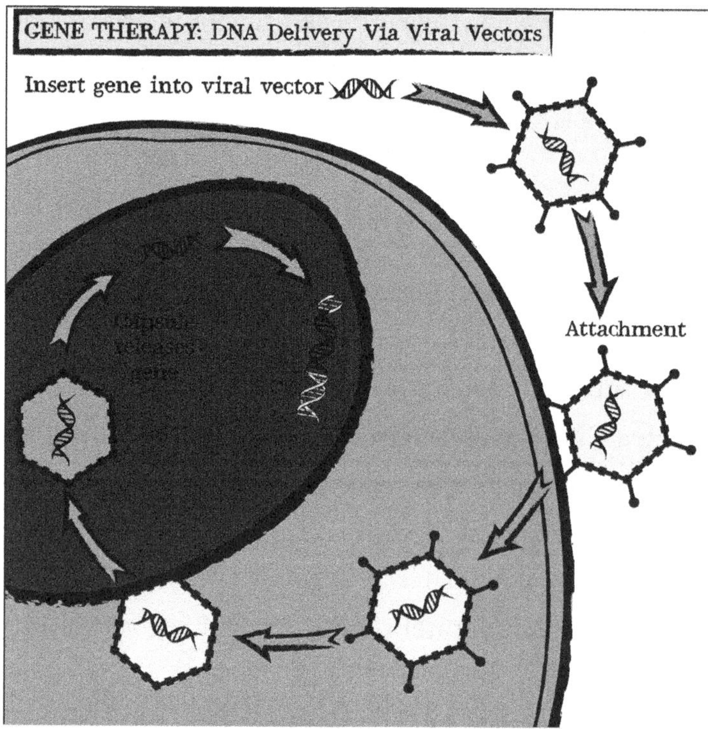

Since Jesse Gelsinger's death, scientists have been especially aware of vector **immunogenicity**, the potential of a substance to induce an immune response. At best, immunogenicity renders a therapy ineffective—if the immune system attacks the vector, its cargo will remain undelivered. At worst, immunogenicity is fatal, as in the case above. Today, gene therapies have been selected for their lack of immunogenicity. However, scientists continue to carefully monitor clinical trials for this response.

GOING FURTHER: *In Vivo* versus *Ex Vivo* Gene Delivery

Viral vectors transport genes in two ways: *ex vivo* or *in vivo* **delivery**. *Ex vivo* delivery involves removing cells from a patient's body, treating them with the vector in the lab, and then re-infusing the changed cells. For example, medical professionals can easily remove blood cells, manipulate, and re-infuse them.

In vivo delivery—targeting cells inside the body—poses considerably more challenges. They include targeting the correct tissue, achieving the proper dose, and avoiding an immune response. *In vivo* delivery takes place via intravenous infusion of the viral vector or direct injection into the target tissue when possible (for example, retinal injection).

Types of Viral Vectors

Given its complexity, it's unsurprising that gene therapy involves more than one type of vector. The most common are **adeno-associated viruses (AAV)**. However, AAVs aren't the only transportation in town. Companies are also investigating **lentiviral vectors**.

Vector choice depends partly on the target tissue and the size of the genetic payload. AAV vectors so far seem to be best at targeting eye, muscle, and central nervous system tissues. Lentiviral vectors are used with cells that divide frequently, such as blood and bone marrow cells. That's because AAV vectors typically can't incorporate therapeutic genes into a host cell genome.

The vector types also differ in the amount of risk they involve. The medical community prefers to used AAV vectors whenever possible because lentiviral vectors pose a slight risk of causing "**insertional mutagenesis.**" This cancer results from the disruption of genes important for regulating cell growth and division. However, cells that divide frequently, such as blood cells, tend to "lose" therapeutic genes. When a cell divides, it copies all of its chromosomes, but not any other cellular DNA. Consequently, those cells need vectors that take the therapeutic gene directly into their genome. Lentiviral vectors fit this bill. Finally, they can carry bigger loads than AAVs, so they're better at moving very large or multiple genes. Lentiviral vectors have thus far only been used to modify blood cells taken out of the body that can be screened before re-infusion to eliminate the risk of insertional mutagenesis. Researchers continue to work on the safety of lentiviral vectors. For example, some companies have designed vectors that allow better control over where in the target genome a therapeutic gene integrates. We expect to see lentiviral vectors being used in more clinical trials as improvements continue.

The European Medicines Agency (EMA) has approved gene therapy treatments for SCID, lipoprotein lipase deficiency—a metabolic disorder—, spinal muscular atrophy and inherited retinal degeneration. In the United States, the FDA approved retinal therapy and treatment for spinal muscular atrophy. Both agencies have also approved CAR-T therapies, one of the immunotherapies discussed in Chapter 4. These are classified as gene therapy because they're created by using lentiviral vectors to engineer T-cells. Let's find out more about Luxturna—the first gene therapy for an inherited disease available in the United States.

Opening Eyes with Luxturna

At the end of 2017, the FDA approved Luxturna for a rare form of inherited blindness: "biallelic RPE65 mutation-associated retinal dystrophy."

- Biallelic: Pertaining to both copies of a particular gene (allele)
- RPE65: A retinal protein that helps convert light into the electric signals the brain interprets as sight
- Retina: Light-sensitive tissue in the eye
- Dystrophy: Wasting of tissue

The corrected RPE65 gene helps repair patients' retinal health. Clinical trial participants' vision loss ranged from mild to severe. The trials included 29 people from ages four to 44. Luxturna improved 93% of patients' eyesight.

Tricky Terms

Vector

In molecular biology, the word **vector** describes a DNA molecule that carries foreign genetic material into a cell. Lentiviral and AAV vectors are one variety. Plasmids are another. These small, circular pieces of DNA come from bacteria. Plasmids and their use in genetic engineering are discussed extensively in *The Biotech Primer One: The Science Driving Biopharma Explained.*

GOING FURTHER: Value-Based Pricing

As one-time, curative treatments, gene therapies will be very expensive. Luxturna, for example, costs $425,000 per eye. Most patients need it for both. The manufacturer is discussing payment options with insurers to help allay the sticker shock. One plan in the works hinges on results. Should the therapy fail to achieve its intended outcomes at certain intervals post-treatment, Luxturna's maker will provide refunds. Another strategy would allow insurers to spread payments over multiple years. The ultimate resolution to the problem of pricing this seminal treatment will likely set an industry-wide precedent.

Looking Ahead

In 2019, the FDA approved a second gene therapy for an inherited disease, spinal muscular atrophy (SMA). We can look forward to the approval of additional gene therapies for diseases caused by a single gene in the coming years. Treatments in the works include those for hemophilia, sickle cell anemia, muscular dystrophy, and other types of hereditary blindness.

Genome Editing: Cellular Self-Help

Gene therapy is revolutionary, but it's not the only genetic approach to curing hereditary illnesses. Rather than delivering a corrected version of a mutated gene to remedy disease, as in gene therapy, **genome or gene editing** supplies the tools afflicted cells need to actually fix faulty genes. Let's take a look at how CRISPR/Cas9 gene editing works, starting with its origins.

BAC Fights Back

Everything we know about CRISPR genome editing we learned from bacteria. Like us, these microorganisms fall prey to viral infection. Their immune systems developed a fascinating way to repel the invaders. In the 1980s, scientists observed a pattern in bacterial genomes: repeating, palindromic sequences, with unique sequences—"spacers"—between repeats. They bestowed a tongue twister of a name, "**c**lustered **r**egularly **i**nterspaced **s**hort **p**alindromic **r**epeats," on the mechanism. What, by the way, is a palindrome? It's a word, phrase, or series of characters (like a DNA sequence) that's the same backward or forward. The word "madam" or phrase "taco cat," for instance.

Everyone just calls the technology CRISPR. At the same time, they discovered the cool genetic repeats, microbiologists noticed that CRISPR sequences always occur near genes that code for an enzyme that cuts DNA. They nicknamed the enzyme Cas, short for "CRISPR-associated."

In the mid-2000s, microbiologists realized these spacers matched the DNA sequences of infecting viruses. The sick bacteria were stashing bits of the offending viral DNA between their own CRISPR sequences! These viral DNA snippets create a "genetic memory," which enables the bacteria to fight back if reinfected. Here's how:

- Viral DNA present in the spacer sequences is copied into viral RNA.
- This new RNA has a portion complementary to the invading viral DNA, and a portion referred to as the Cas binding scaffold.
- The bacterium makes the DNA-cutting enzyme Cas.
- Cas recognizes and binds to the binding scaffold portion of the new viral RNA.
- The resulting viral RNA/Cas complex finds its match on the invading viral DNA.

- The RNA attaches to the DNA. The Cas enzyme cuts up the foreign DNA, destroying the virus.
- Voila—“healthy” bacteria.

In 2012, researchers adapted this defense for human cells. Molecular biologists realized they could design and synthesize RNA sequences that contained a portion complementary to a specific human gene sequence and a Cas binding scaffold. These are referred to as “guide RNA.” When a guide RNA and Cas enzyme are introduced into a human cell, the guide RNA directs the Cas enzyme to cut DNA in a sequence-specific location, just as it would in bacteria. This original Cas protein came from the Cas9 *Streptococcus* bacteria—hence the moniker CRISPR/Cas9.

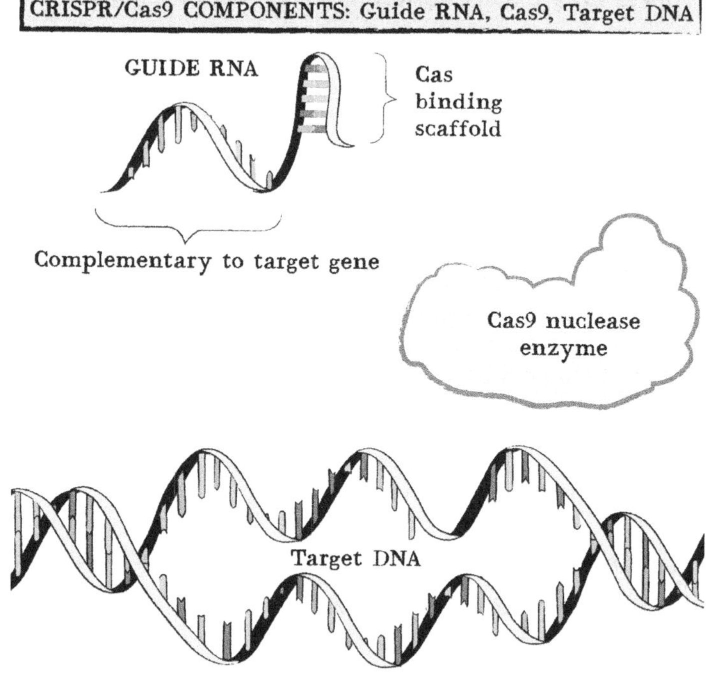

Specifically, Cas9 creates double-stranded breaks in the DNA—it severs *both* strands of the double helix. The secret to genome editing lies in these double-stranded breaks in strategic locations within the genome. What happens after those breaks occur is biotech magic. There are two possibilities: 1) disrupting a disease-causing gene, so it no longer makes the disease-associated protein, or 2) using a repair template to overhaul the disease-causing gene to display the correct sequence.

Think of DNA as a two-lane bridge. Imagine an earthquake that causes a span to rupture and fall away. The gap represents a double-stranded break. The road crew (in this case, cellular enzymes that repair DNA) can fix the break in two ways:

- Non-Homologous End-Joining (NHEJ): Visualize workers pushing the two remaining sections of the bridge back together. NHEJ can result in a sequence error, just as sections of a bridge can line up improperly. Because the repair occurs in the middle of a gene, it typically disrupts gene function and the production of the corresponding protein.
- Homology Directed Repair (HDR): This approach relies on a highly similar (homologous) DNA segment to fix the break. In the case of our earthquake-stricken DNA bridge, workers build the new section offsite and use it to mend the bridge. This new section is referred to as the repair template.

Typically, a DSB activates an NHEJ repair. This disrupts the faulty gene. This benefits a patient if a gene produces a protein that causes his disease. Note that this genetic repair doesn't *correct* a faulty gene.

To *fix* a gene sequence—change a mutated base (A, C, G, or T) to the correct one—a "repair template" must be delivered with the break, activating the HDR pathway. This repair template is virtually identical to the section of DNA being edited. It differs only in that it contains the right base or bases for the damaged gene.

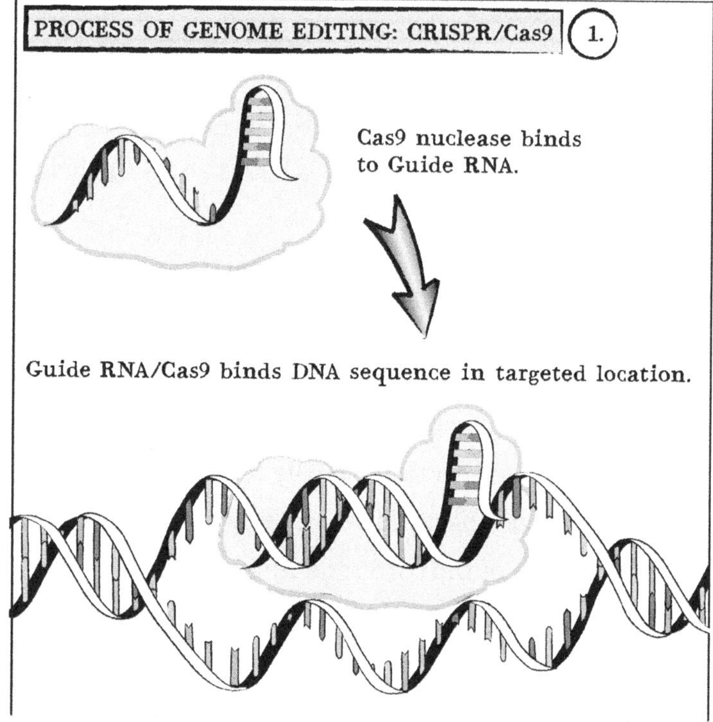

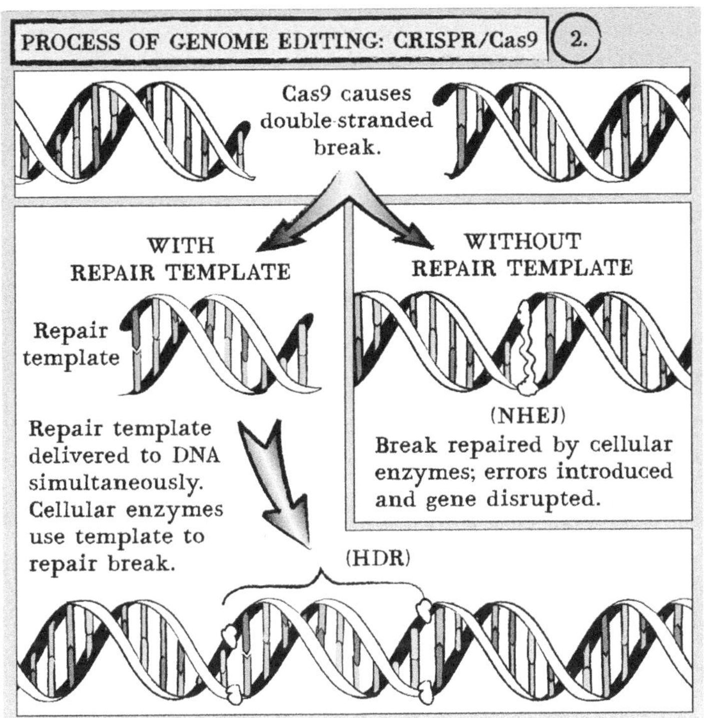

INDUSTRY NOTE: Clinical Applications of CRISPR

CRISPR genome editing is currently in a few of clinical studies. The first treatment to enter clinical trials involved knocking out a gene called PD-1 in melanoma, synovial sarcoma, and multiple myeloma patients' T-cells. Because PD-1 inhibits T-cell activation, taking it out allows the immune cells to activate more fully. This, in turn, enables patients' immune systems to better fight cancer. This treatment is administered *ex vivo*—removing the patients' T-cells, modifying them in the lab, and reinfusing them.

Phase I/II CRISPR trials are also underway for an inherited retinal disease. Leber congenital amaurosis is caused by an error in a gene necessary to develop retinal cells, CEP290. The mutation changes how the retina processes the gene's RNA so that it can't manufacture CEP290 protein. By using CRISPR to delete the mutated base, researchers hope to restore the protein production and hence retinal function. This treatment is the first genome editing treatment patients receive directly, or *in vivo*, by injection.

GOING FURTHER: INDIRECT EFFECTS

In some cases, disrupting one gene enhances the expression of another. Researchers are exploiting this effect to develop a genome-editing treatment for a rare blood disease, beta-thalassemia. The malady, which reduces the hemoglobin production, results from a mutation in the hemoglobin gene. (Remember, the protein hemoglobin transports oxygen in our blood.) Genome-editing companies are now testing CRISPR-based treatments that knock out the production of a related gene called BCL11A. This genetic cousin makes a protein that suppresses the production of fetal hemoglobin, which normally only babies make. "Tricking" red blood cells into continuing to generate fetal hemoglobin may enable patients to compensate for their deficit in the adult version.

GOING FURTHER: CRISPR/Cas9 Delivery

For CRISPR/Cas9 components to do their job, they need to get to the nucleus of their target cells. How though? As in the case of gene therapy, doctors deliver the gene for the Cas9 protein and guide RNA by viral vector. Then the cell makes the Cas9 and guide RNA themselves.

New Cas on the Block

While Cas9 is the most well-known CRISPR enzyme, others have been developed from related bacterial proteins. They have slightly different applications. Two of these are RCas9 and Cas13.

The RNA Editor

Researchers at the University of California, San Diego (UCSD) developed RCas9, a modified version of Cas9. RCas9 targets messenger RNA (mRNA), the intermediary between DNA and proteins. Recall from Chapter Two that information stored in genes is converted into mRNA. Cells use mRNA to make a protein. In other words, mRNA serves as a temporary copy of the permanent information DNA stores.

RCas9 cuts mRNA. Scientists can target specific disease-associated mRNAs with RCas9. The UCSD team has tested RCas9 on cell-based disease models with encouraging results. They've shown that RCas9 can reduce problematic mRNAs in models of Huntington's disease, myotonic dystrophy, and amyotrophic lateral sclerosis. Focusing on disease-associated RNAs

Tricky Terms

Knockouts and "Knock-ins"

When scientists disrupt a specific gene in a cell or an organism, they refer to the procedure as "**knocking out**" that gene. This kind of genome editing is only beginning to be used on people. However, scientists have used it in lab research for years. This technique enables them to create "**knockout mice**" that contain a specific disrupted gene. The modified mice help researchers understand a particular gene's function does by observing what happens in its absence. They also provide clues about the physiological effects of a drug that strongly inhibits a specific protein.

Knock-in refers to the insertion of a specific DNA sequence into a cell line or organism. Knock-in mice can provide an accurate model of human disease. Molecular biologists have created mice expressing human genes related to cancer, Alzheimer's, diabetes, and other health problems. In preclinical drug development, these mice may provide better evidence than unmodified animals as to whether a drug will work in humans.

Prior to CRISPR/Cas9, researchers used other technology to create mouse models, but the process required months or years. With CRISPR, the mice making takes just a few weeks.

instead of genes may allow researchers to treat certain genetic diseases by getting at their root cause without permanently altering someone's genome.

A New Model of Genetic Scissors: Cas13

Researchers at the Broad Institute in Cambridge, MA, recently discovered another Cas enzyme, Cas13. This enzyme is doubly useful. *Unmodified* Cas 13 cuts up RNA that codes for a specific protein. This "RNA knockdown" destroys or limits the protein for which it codes. It can be useful in research, as described above, for knockout mice. It also holds the potential to reduce disease-causing proteins.

A *modified* version of Cas13 serves as a base editor. That's because it edits single bases that comprise the mRNA sequence instead of only cutting mRNA. Specifically, the new and improved Cas13 changes the base adenine (A) to G). Mutations that transform G to A play a role in Duchenne muscular dystrophy and Parkinson's disease. This fact suggests that Cas13 may help treat these and perhaps other diseases by restoring a patient's normal protein function. As with Cas9, the ability to do so without permanently modifying the genome may also make it desirable for certain applications. The Broad Institute team christened the modified Cas13 enzyme **REPAIR**, or RNA Editing for Programmable A to I Replacement.

Genome Detectives

Scientists have also begun using CRISPR technology for new diagnostics. Unmodified Cas13 features another attribute that makes it a powerfully diagnostic tool. Once activated, it chops up not just its target mRNA but also any other mRNA nearby. Researchers call this "collateral cleavage," but you can think of it as an enzymatic RNA feeding frenzy. Let's see how this feeding frenzy is put to use as a diagnostic.

Many viruses have RNA-based genomes, including Zika and HIV. Broad Institute scientists have developed diagnostics that consist of Cas13 combined with guide RNAs that target a virus-associated sequence. If a sample contains the sequence of interest, Cas13 finds and cuts it. How to recognize when this happens? Enter reporter RNAs. These RNAs contain a fluorescent tag that's released and becomes visible *only* if the reporter RNA is cut. In other words, fluorescence appears only if the Cas13 enzyme finds its target and is activated to cut the target as well as any other RNA in sight. The Broad team calls this system "Specific High-sensitivity Enzymatic Reporter unLOCKing"—SHERLOCK. It's been incorporated into a "lateral flow assay" for Zika. This diagnostic comes on a stick, similar to a pregnancy test.

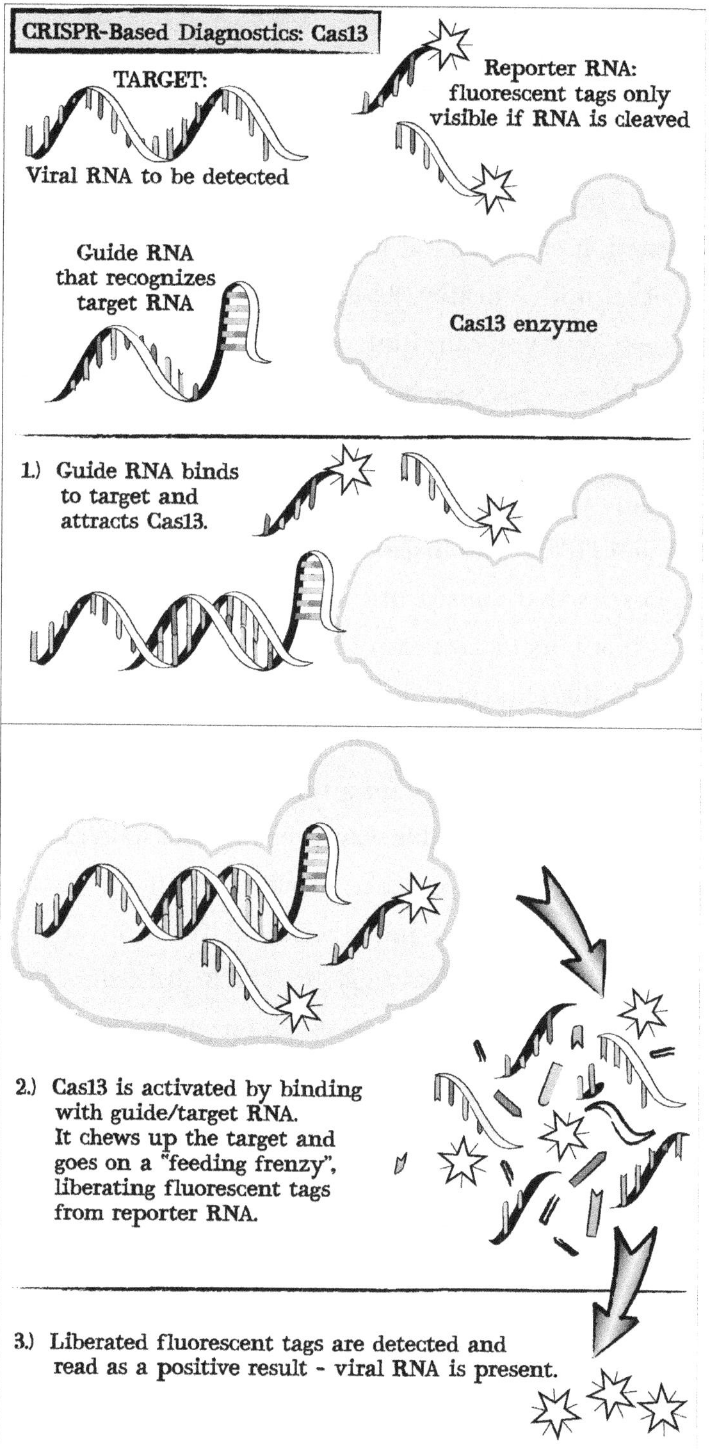
CRISPR-Based Diagnostics: Cas13
TARGET:
Viral RNA to be detected
Reporter RNA: fluorescent tags only visible if RNA is cleaved
Guide RNA that recognizes target RNA
Cas13 enzyme
1.) Guide RNA binds to target and attracts Cas13.
2.) Cas13 is activated by binding with guide/target RNA. It chews up the target and goes on a "feeding frenzy", liberating fluorescent tags from reporter RNA.
3.) Liberated fluorescent tags are detected and read as a positive result - viral RNA is present.

SHERLOCK'S Counterpart: DETECTR

UC Berkeley scientists have developed a genomic detective of their own: DETECTR. This diagnostic, DNA Endonuclease Targeted CRISPR TransReporter, works with yet another Cas enzyme, Cas12a. This enzyme also cuts a specific DNA sequence with the help of guide RNA. Like Cas13, it then attacks nearby DNA sequences. Its destruction includes unleashing fluorescently-labeled DNA reporter segments. The Berkeley team has demonstrated DETECTR's ability to identify the human papillomavirus (HPV) in blood samples. DETECTR should also be able to detect cancer and other disease-associated mutations in DNA samples.

CRISPR-based diagnostics promise two important advantages over other kinds of tests: extreme sensitivity—the ability to correctly identify people who have a disease—and greater specificity. This is the ability to generate a negative result for those who *don't* have it. The precision arises from CRISPR-based diagnostics' capacity to detect minute amounts of DNA or RNA. Not only that, they can identify disease based on a specified gene sequence. This potent combination of attributes lays the groundwork for diagnosing infectious diseases and cancers more quickly and accurately.

Challenges to Genetic Manipulation

As cutting-edge therapies, gene transfer and genome editing hold a lot of promise but carry unknowns and potential pitfalls as well. These include:

- Specificity: Ensuring that genome editing occurs only at specific, intended sites within the genome. One wrong cut could cause serious damage.
- Efficiency: Efficiencies of gene deletion, insertion, and/or replacement may not be high enough to drive therapeutic benefit, especially for *in-vivo* delivery of new genes or gene editing tools.
- Immunogenicity: One of the challenges to developing viral vectors for gene therapy is vector immunogenicity. Researchers select vectors partly based on how they affect our immune system. For genome editing, the concern applies not only to the viral vector that delivers the editing tools but also the nuclease protein as well which may also induce immunogenicity.

Treating hereditary diseases by modifying the genome is finally becoming a reality after decades of effort by academic and industry-affiliated scientists. In the next chapter, we look at another new way to modify gene expression: nucleic acid-based therapies.

CHAPTER 7

RNA-Targeted Therapeutics

Sometimes it feels biopharma is awash in an alphabet sea: DNA, mAbs, AAVs, Cas...While a vast array of acronyms does accompany this amazing technology, any discussion of the field would be incomplete without covering another three letters: RNA. Specifically, this chapter is going to explore the mechanisms that allow researchers to use RNA as medicine or develop medicines that take advantage of these molecules' actions.

First, let's start with RNA itself. There are several different types of RNA, but the one most are familiar with is messenger RNA (mRNA). Reviewed extensively in *The Biotech Primer One: The Science Driving Biopharma Explained*, mRNA serves as the template for the production of proteins. This template is often called a transcript because it is a copy of the information found in genes. In the 1970s, molecular biologists discovered that cells also produce what they referred to as **antisense** RNA transcripts (see Tricky Terminology: Antisense). These transcripts resulted in *reduced* production of specific proteins. This

discovery set the stage to develop an entirely new class of drugs that could work with a naturally-occurring cellular enzyme to block the production of disease-associated proteins.

Since then, the FDA has approved a handful of drugs utilizing this strategy. More than a hundred RNA-targeting drugs are currently in clinical development for health conditions from neurodegeneration and metabolic disorders to various cancers and infectious diseases. Because they work at the genetic level, RNA-targeted therapeutics make potent tools that can potentially address diseases that are otherwise untreatable with traditional small molecule or biologic drugs.

But first, what is an antisense RNA? Unlike mRNA, which codes for proteins, these short pieces of RNA have a sequence that is complementary to mRNA. As a result, the antisense strand binds to the mRNA strand. This connection creates a piece of double-stranded RNA. Different cellular enzymes recognize and destroy double-stranded RNA. Thus, the mRNA gets destroyed. No mRNA, no protein. No protein, no disease.

Tricky Terms

Antisense

The term "antisense" derives from the fact that the physical structure of DNA and RNA necessitates that cellular enzymes read them in a specific order. The enzymes that read mRNA and convert it to proteins process it from the five prime (5') end—the front end of the molecule—towards the three prime (3') end—the back end. The chemical configuration of the A, C, G, and T bases that make up RNA dictates this orientation. The "sense" orientation is reading the molecule from 5' to 3.' The antisense orientation reads in the opposite direction. An antisense RNA is complementary and has the opposite orientation of a protein-coding mRNA. To illustrate,

mRNA:
5'ACGGATTCAGTA3'

Antisense:
3'TGCCTAAGTCAT5'

Regulatory RNA

Since the discovery of antisense RNA transcripts, scientists have found other types of RNA that modulate gene expression. Collectively, these are "regulatory RNA." Molecular biologists are still deciphering the different kinds and their functions. Two of the best characterized are antisense RNA and short interfering RNA (siRNA). Molecular biologists have figured out ways to mimic their actions to treat illness. They do this by creating synthetic versions of the regulatory RNA to modulate the expression of a disease-associated gene.

GOING FURTHER: Synthetic RNA and DNA

To make synthetic RNA or DNA, a technician programs the correct sequence of nucleotide bases (As, Cs, Gs, and Ts) into a machine which then assembles them through chemical reactions and tah dah—short sequences of nucleotide bases materialize. This process is sometimes referred to as "printing DNA/RNA."

Antisense

Antisense drugs are short (typically 20 to 50 bases long) synthetic pieces of antisense RNA. Molecular biologists design them to be complementary to the sequence of a disease-associated mRNA. In other words, they're targeting an mRNA that codes for a disease-associated protein. The antisense drug is administered to the patient via injection. When it enters the target cell (more on that in the Going Further: Delivery box), the drug forms a double-stranded RNA complex with its complementary target. This triggers an enzyme called RNAse H, which destroys the antisense-mRNA duo. Without the mRNA, our bodies can't produce the protein. Antisense drugs target and destroy disease-associated mRNAs, which stops the affliction in its tracks.

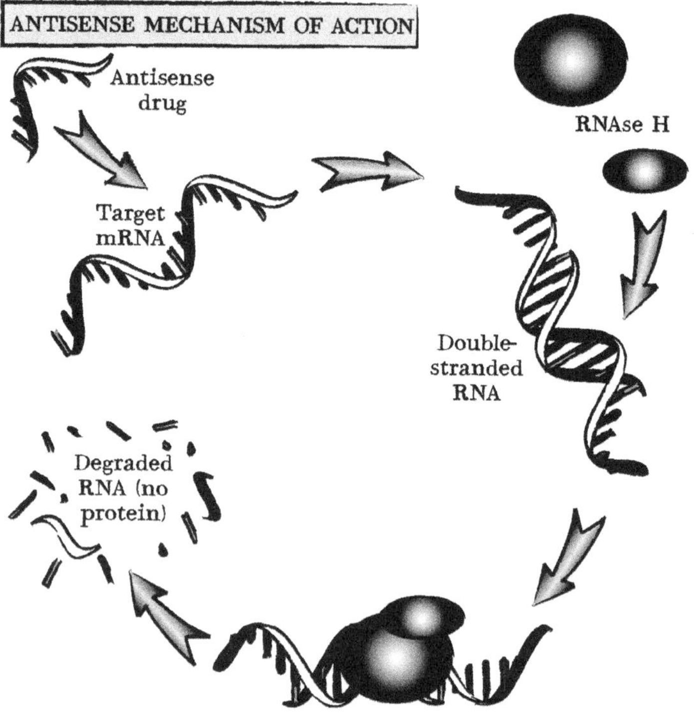

The drug Kynamro makes an excellent illustration. This drug treats familial hypercholesterolemia, an inherited form of severe high cholesterol. The medicine reduces the amount apolipoprotein B—a major component of LDL cholesterol—in a patient's blood. This, in turn, lowers cholesterol levels.

GOING FURTHER: Delivery

As in the case of protein therapeutics, people can't just pop a synthetic RNA "pill." If taken orally, a person's digestive tract would break the drug down before it could be absorbed. Instead, patients receive antisense drugs by injection. Once in the bloodstream, they must then enter their target tissue.

Over the years, ensuring that RNA-based therapies reach their target tissue has been a challenge for drug developers. To get around this obstacle, some have encapsulated the medicines in lipid nanoparticles and "decorated" those nanoparticles with molecules that have an affinity for a target tissue. Manufacturers also take advantage of the fact that certain parts of the body make better targets for RNA therapies than others. That's because the antisense drugs accumulate in particular spots such as the liver, kidney, spleen, bone marrow, and fat cells. Diseases that affect these tissues make good candidates for treatment with RNA-based therapeutics. Doctors also consider the eyes and lungs likely prospects because it's relatively easy to deliver these drugs through eye drops or inhalants.

Meet Introns and Exons

Antisense technology continues to evolve. Recently-approved drugs such as Exondys 51 and Spinraza alter how cells process "pre-mRNA." This long strand of RNA results from cellular enzymes transcribing a gene. Before the pre-mRNA can be used to make proteins, it must first be processed by splicing enzymes in the cell. These enzymes cut or add specific pieces before the pre-RNA matures into the mRNA that transmits the protein-making instructions to the ribosomes.

Pre-mRNA initially contains two sections—**introns** and **exons**:

- Introns get cut from pre-MRNA to make mature mRNA.
- Exons are the regions that remain in mature mRNA.

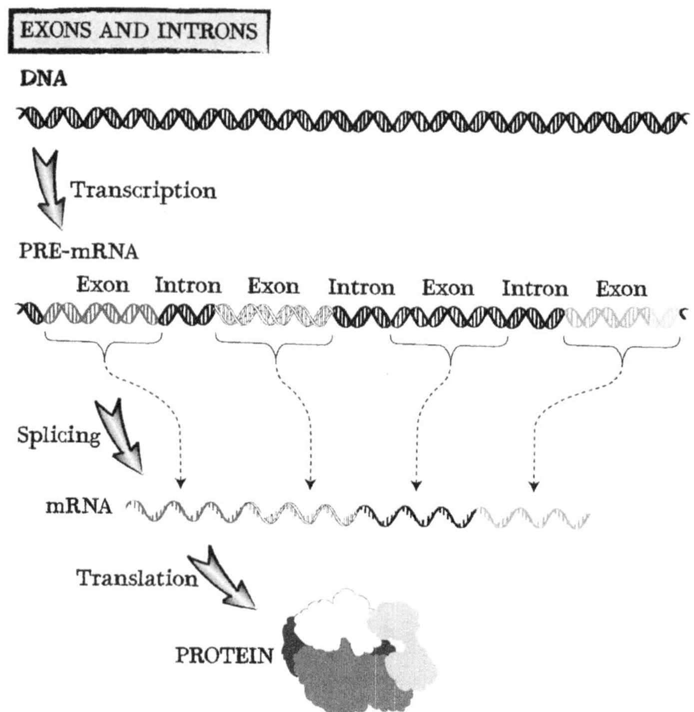

Drug developers have begun to explore a new approach to some genetic diseases by changing how affected cells process pre-mRNA. One example is Duchenne's muscular dystrophy (DMD). This condition causes progressive muscle degeneration and weakness. It can now be treated with an antisense drug that changes pre-mRNA processing.

DMD arises from a mutation in the gene that codes for the muscle protein dystrophin. The protein forms part of a larger group that strengthens and protects muscle. Faulty dystrophin leads to damaged and progressively weaker muscles. Kids with DMD display developmental delays almost from birth. Over time, standing and walking become impossible. Teenagers and young adults often develop cardiomyopathy or fatal respiratory problems because the muscles that support breathing deteriorate so much. The drug Exondys 51 binds to the exon containing the mutation so that it stays out of the dystrophin mRNA. This allows DMD patients to produce a more functional dystrophin protein. The drug's action exemplifies a technique referred to as exon-skipping

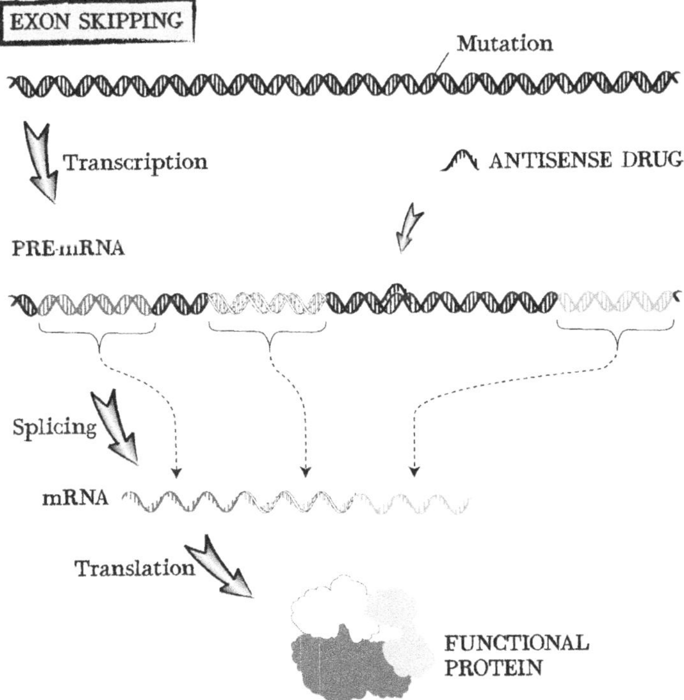

GOING FURTHER: Inspiration for Exon-Skipping

Simply removing a mutated exon from a protein wouldn't result in a more functional protein in most cases. Sometimes however, the stars align. The inspiration for Exondys 51came from patients with a similar condition called Becker muscular dystrophy. They naturally produce a truncated dystrophin protein and experience a less severe symptoms. This difference led molecular biologists to wonder what affect an artificially shortened version of the protein would have on kids suffering from DMD.

Short Interfering RNA

Like antisense, short interfering RNA (siRNA) is produced naturally by cells to reduce the translation of target mRNAs into their corresponding protein. Drug discovery researchers have created synthetic siRNAs to target disease-associated proteins. siRNA differs from antisense in two ways:

- It starts out as double-stranded rather than single-stranded.
- It's processed by different cellular enzymes (DICER and RISC rather than RNAse H)

Synthetic siRNA molecules start out as "short hairpin RNA." The cellular enzyme Dicer processes them to become siRNA. "Hairpin" refers to the fact that these bits of RNA look like old-fashioned hairpins, with a loop at the top:

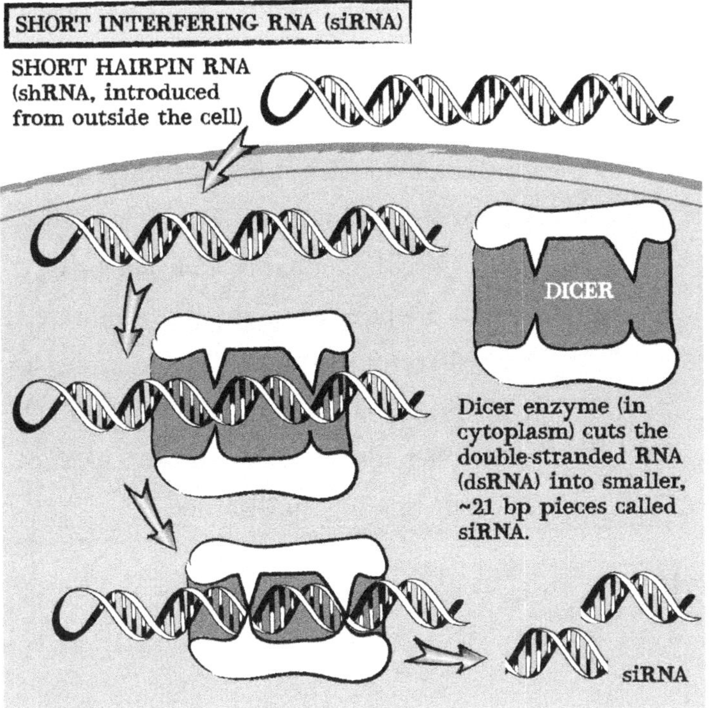

Next, the siRNA binds a second enzyme, RISC (RNA-induced silencing complex), which releases one of the two strands of siRNA. The remaining strand is complementary to the target mRNA. Scientists refer to it as the "guide strand" because it directs the target mRNA into position within the RISC complex:

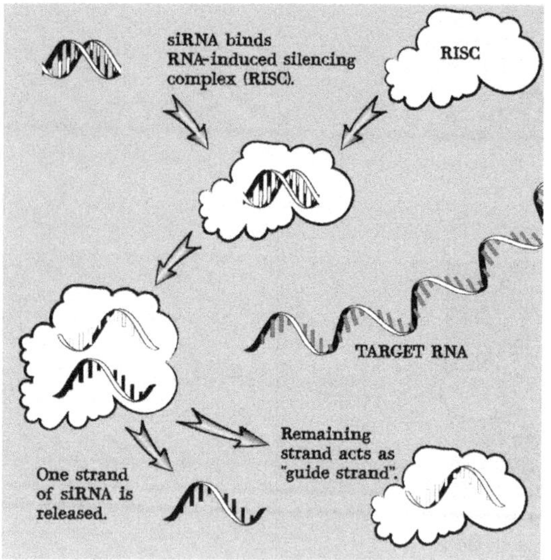

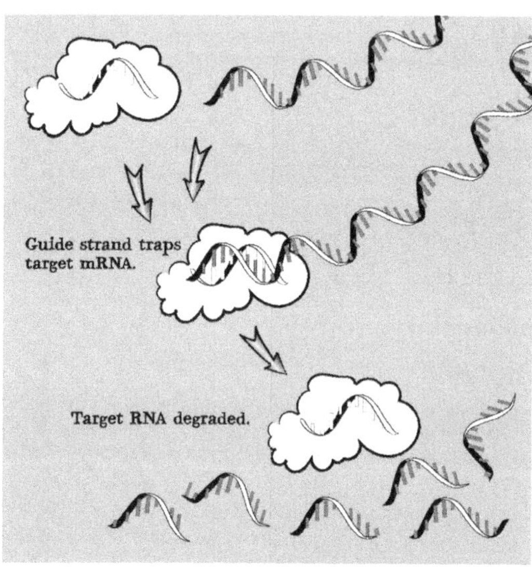

When the guide RNA traps the target mRNA, RISC degrades it

The first siRNA drug, Patisiran, came to market in 2018 for hereditary transthyretin amyloidosis. This condition causes cardiomyopathy and neuropathy due to the buildup of clumps of mutated transthyretin protein in cardiac muscle and nerves. Patisiran destroys the mRNA that codes for the mutated protein, successfully treating the disorder.

GOING FURTHER: The Regulome

Thanks to the Human Genome Project, we know that our genome contains about 20,000 genes—far fewer than the originally predicted 100,000. Surprisingly, these genes only make up about two percent of the genome. Most of the genome lies outside of the coding regions. Geneticists once described this non-coding DNA as "junk DNA" because it served no apparent purpose. More recently though, they've uncovered evidence that about 85 percent of the non-coding region is transcribed into "regulatory RNA." Scientists sometimes call this portion the **regulome**. Our increased understanding of the regulome has led to the emergence of RNA-targeting drugs.

Industry Focus: mRNA Drugs

This chapter describes drugs that target mRNA, destroying it to prevent it from becoming a disease-associated protein. We now turn our attention to drugs that *are* mRNA.

Some strands of mRNA encode medically useful proteins. Once inside the patient, her cells render the mRNA into a therapeutic protein. No mRNA drugs have made it to the approval stage yet, but several are in the pipeline.

The farthest along are mRNA-based vaccines being developed as cancer immunotherapies. They draw on the idea of neoantigen vaccines described in Chapter 5. Researchers identify a protein that is unique to a particular cancer and use it as the basis of a therapeutic vaccine to activate the patient's immune response. For mRNA-based neoantigen vaccines, the vaccine itself would contain mRNA encoding the neoantigen, rather than the protein itself.

When fully realized, mRNA drugs will essentially be biologic drugs that our own cells make. If successful, one key advantage over existing biologics would be the mRNA drugs' ability to introduce therapeutic proteins into a patient's cells. Because current biologics are too big to enter cells, they consequently work by interacting with proteins on the cell surface or in the bloodstream. Manufacturing costs are also lower for mRNA than for proteins.

This chapter explores an emerging category of drugs, RNA-targeted therapeutics designed to change gene expression. In the next chapter, we'll dive into a whole new world: the microbiome.

CHAPTER 8

The Microbiome: Disease And Cure

At first glance, it might seem as if the hundred trillion microbes that reside in and on us have little to do with biopharmaceuticals. But as previous chapters have illustrated. Bacteria, viruses, and other microorganisms have "big" roles to play in human health. Let's see how.

Over the past decade, microbiologists have come to realize that the countless microbes living on and inside of us play a critical role in our health. These organisms: bacteria, fungi, viruses, and protozoa, collectively comprise our **microbiome**. We've long understood that the microbiome profoundly influences our digestive health. Now, however, we've learned it affects many health conditions, including allergies and asthma, autism, Parkinson's disease, depression and anxiety, obesity, and diabetes.

You'll Never Walk Alone

Our every surface and crevice teem with microbes. In fact, bacterial cells in and on our body outnumber human cells by about ten to one. Our cells are much larger than bacteria cells, however, so we're still mostly human. People generally walk around completely unaware of their invisible passengers. However, these little hitchhikers form an integral part of what makes us both who and *how* we are. This symbiotic relationship has existed for eons.

Until fairly recently, the pharmaceutical industry pretty much ignored the wild micro kingdom. Doctors have largely concerned themselves with the minute percentage of bacteria and other microbes that cause illnesses such as pneumonia, tuberculosis, and meningitis. Not anymore. We're finding out more and more about how our personal ecosystem helps regulate processes such as digestion and immunity and keeps disease-causing bacteria in check.

What does this evolving perspective mean for biopharma? Experimental treatments involving fecal transplants (known as bacteriotherapy) to restore a healthy gut microbiome have shown promise for relieving inflammatory bowel disease. However, these approaches are unlikely to become mainstream due to the lack of standardization, discomfort, and marketing issues.

Microbiome-Based Therapy: More Than a Gut Feeling

Cocktail Fodder

For every 100 pounds a person weighs, bacteria constitute about two pounds.

The premise of microbiome-based therapeutics is straightforward. First, identify how the microbiome of a person with a disease differs from a healthy person's. Next, try to make the impaired microbiome more like the healthy one. Drug discovery researchers have narrowed in on a few strategies. They range in scale from transplanting fecal matter to identifying specific beneficial bacterial strains to pinpointing specific bioactive compounds secreted by microbiota.

Most of our microbiome lives in our gastrointestinal tract—this is the gut microbiome. Fecal matter transplant (FMT) completely resets a patient's gut microbiome by transferring part of the microbe-rich community from a healthy donor through a colonoscopy or enema into a patient.

Feces as "medicine" in any capacity sounds a bit suspect. Surprisingly though, the idea has been around for over a thousand years. Chinese health practitioners administered a form of FMT to patients with life-threatening diarrhea as long ago as the fourth century, according to ancient medical texts.

In the past decade, clinical trials have suggested that FMTs effectively treat recurrent, antibiotic-resistant *C. difficile*. This serious bacterial infection causes severe diarrhea and sometimes inflammation of the colon. Bacteriotherapy also shows promise for other problems such as inflammatory bowel disease. Experiments in mice suggest that transferring gut **microbiota** from one mouse into another impacts metabolism, turning a skinny mouse obese and vice-versa.

How do scientists think FMTs impart a therapeutic benefit? With *C. difficile* infection, it's thought that most of us already harbor low levels of bacteria. Antibiotic use, however, sometimes wipes out the beneficial bacteria that keep *C. difficile* under control. Repopulating a patient's gut via FMT restores the proper balance, reigning in *C. difficile*. The improvement of symptoms seen in inflammatory bowel disease patients after FMT likely relates to the regulation of our immune system. About 75% of human immune tissues reside in our digestive tract. Consequently, immune cells interact intimately with gut microbiota. Changes to the gut microbiome can influence how immune cells behave, potentially resulting in inflammation.

FMT Standardization

One obstacle to FMT becoming routine therapy is the lack of standardization. Biotech companies are addressing that by attempting to create standardized microbiota suspensions of stool samples from rigorously screened, healthy donors. Some are currently in clinical studies for recurrent *C. difficile* infection, vancomycin-resistant *enterococci*, pediatric ulcerative colitis, multi-drug resistant urinary tract infections, and hepatic encephalopathy. Some drug developers are also working on a more user-friendly transplant–a capsule. This less invasive alternative is now in clinical studies for recurrent C. *difficile* infection.

Tricky Terms

Monoclonal Microbial
A **monoclonal microbial** is a specific strain of bacteria with proven therapeutic benefits.

Strength in Numbers

A second approach to transferring the benefits of the microbiome comes in the consortium approach. Here, researchers look at specific bacterial strains or communities of strains (i.e., consortia) with likely medical applications. These "bugs" are then isolated to develop into standardized therapies. In broad strokes, here's how:

- Collect clinical data from studies that investigate the human gut microbiome.
- Identify bacterial communities associated with good clinical responses.
- Test them and select those with the most potent pharmacological affects.
- Assemble the selected strains in consortia that work well together.
- Use the strains to make Good Manufacturing Practices (GMP)-grade drugs for clinical studies.

A variety of consortia-based products are currently in clinical development.

INDUSTRY NOTE: Bacterial Byproducts

Other scientists are zeroing in even closer, microbiomically speaking, by trying to isolate beneficial bacterial byproducts. Some of the proteins or small molecules secreted by bacteria pack a therapeutic punch. This approach most closely resembles traditional drug discovery. This means that potential microbiome-derived drugs will travel a well-defined pathway toward clinical development and manufacture. This makes them very attractive to biotech companies. However, many scientists believe that most beneficial effects derived from the microbiome come from the interaction of different bacterial strains and their byproducts with each other.

GOING FURTHER: The C-Section Connection?

Researchers are busy exploring the health benefits that our own personal zoos may someday yield. However, some research provokes more controversy than others. For example, studies show that babies delivered by C-section have about 20% higher rates of asthma than those born vaginally. C-section babies are also significantly more likely to develop allergies than their traditionally-born peers. It's now thought that the higher risk of these health problems may stem from the fact that C-section births deprive newborns of the protective bacteria that comes from passing through the birth canal.

What's a mother who must undergo a C-section to do about her child's microbiota? Enter a practice called vaginal seeding. In this procedure, the obstetrician transfers some of the mother's microbiota (found in her vaginal fluids) on her child's mouth, nose, and skin using a cotton swab or gauze. Preliminary studies suggest that this transfer endows C-section newborns with some of their mother's protective microbiome. Whether the post-birth swipe translates into long-term benefits remains to be seen. There are also potential risks involved, so the procedure is still considered experimental.

A Closer Look: The Gut-Brain Axis

The latest research on the gut microbiome points to a surprising connection with brain health. Studies at the Flanders Institute for Biotechnology revealed that two strains of bacteria, *Coprococcus* and *Dialister*, occur at low levels in the gut microbiome of people with depression. This early-stage work may one day provide the basis for new antidepressants.

Work done at the California Institute of Technology suggests a possible role for the gut microbiome in Parkinson's disease (PD), a chronic, progressive movement disorder. Symptoms include tremors, slow movement, rigidity of the limbs and trunk, and impaired balance and coordination. These problems arise from the malfunction and death of neurons that produce the neurotransmitter dopamine.

PD affects nearly one million adults in the U.S. Its cause remains unknown. About 75% of PD patients experience gastrointestinal issues such as constipation. This specific symptom provided the impetus to examine a possible connection between gut health and the disease.

The protein alpha-synuclein (αSyn) aggregates in both brain and gut cells in PD patients. Researchers studied a strain of mice that overproduce αSyn. One nest (yes, that's the name for a group of mice) was bred in a completely sterile environment to create germ-free αSyn mice—in other words, without a microbiome. The other αSyn mice grew up in a typical lab environment and developed the normal collection of gut bacteria. On tests to assess motor skills, the germ-free αSyn mice performed significantly better than their typically germy peers.

The result suggests that even in mice that overproduce the αSyn protein, Parkinson's disease depends on the presence of certain microbes. Additional experiments with the germ-free mice suggest that a molecule called butyrate, produced by certain gut bacteria, enters the brain and activates an immune response that damages or kills neurons.

Further evidence supports the gut-brain axis hypothesis. In collaboration with Rush University (Chicago, IL) gastroenterologists, the Cal Tech researchers transferred fecal samples from PD patients into the germ-free αSyn mice. The procedure reset the gut microbiome of the recipient mice so that it matched the patients'. After transplantation, the mice began showing PD symptoms. Transfer of fecal matter from healthy people didn't trigger the same reaction. The experiments suggest that the gut microbiome plays a major part in Parkinson's disease.

These promising findings still need to make their way into human therapeutics. This may be easier than traditional neurological approaches to treating Parkinson's disease. That's because delivering drugs to the gut poses far less challenge than getting them across the blood-brain barrier.

The emerging work on our microbiome puts on new twist on the old expression "think with your gut." As the story unfolds, we are likely to see new therapeutics based on restoring the established balance that arose from millions of years of our co-evolution with microbes.

The next and final chapter will present a more in-depth look at how some of the therapies described earlier are successfully treating disease.

CHAPTER 9
Therapeutic Choices

Today, physicians wield a vast assembly of pharmaceutical tools against disease, from old-school antibiotics to our very own RNA. With even more on the developmental horizon, it can be hard to keep up. In this chapter, we'll walk through some specific examples of disease and their treatments.

Infectious Disease and Its Foes

Starting in the 20th century, doctors have typically treated infectious diseases with small molecule drugs. These include most antibiotics, which fight bacterial infections, and anti-virals.

Antibiotics: The First Line of Defense

People originally discovered many antibiotics as natural substances produced by fungi or other organisms as a defense against bacteria. Penicillin serves as the classic example. Microbiologist Alexander Fleming discovered it in 1928 when he noticed that bacteria didn't grow on his cultures in areas where the *Penicillium fungi* had colonized.

Penicillin belongs to a group of antibiotics called β-lactams. They work by preventing bacteria from making **cell walls**. The cell wall is a vital structural layer composed of proteins and sugars that surround bacterial cells immediately outside their membranes. Without them, bacteria can't survive. Human cells have no cell walls, so penicillin doesn't harm them.

Ever "Better" Bacteria

Initially, antibiotics make short work of bacteria. Unfortunately, most lose their ability to wipe out infection over time because bacteria eventually develop resistance.

Here's an example of just how bacteria become resistant to drugs: Suppose an antibiotic inhibits an enzyme that bacteria need to reproduce. If just *one* bacterium mutates, so the enzyme changes shape just a bit, the antibiotic longer recognizes and blocks the mutated enzyme. The mutant microbe survives and replicates, even as the bacteria susceptible to the antibiotic die off. Over time, this "new and improved" strain becomes dominant. Just like in the bad old days before penicillin, the revitalized infection again spreads from person to person. Sometimes a strain of bacteria develops resistance to several antibiotics. These are multi-drug resistant (MDR) bacteria.

Ever Better Antibiotics

To combat increasing resistance, pharmaceutical companies are developing antibiotics that rely on novel mechanisms. For example, a small molecule drug called gepotidacin inhibits an enzyme that bacteria need to replicate their DNA. No DNA, no more bacteria. Gepotidacin is currently in Phase II clinical testing. Iclaprim is another innovative antibiotic in clinical development. This newcomer targets the infamous MRSA (methicillin-resistant *Staphylococcus aureus*), one of the most problematic infections in hospitals. Iclaprim inhibits an enzyme vital to bacterial metabolism.

GOING FURTHER: Treasure in the Soil, the Oceans

Scientists have traditionally found antibiotics by screening soil bacteria for compounds that they secrete which kill other bacteria that live in the soil and compete for resources. This methodology, however, has a near-fatal flaw. Most of these organisms grow so poorly in the lab that it's impossible to investigate 99% of candidates.

Microbiologists at Northeastern University came up with a method of cultivating bacteria by sandwiching them between layers of soil separated by a semi-permeable membrane. This allowed them to screen an impressive 50,000 then previously untested types of soil bacteria. The innovative effort yielded 25 new drug candidates. So far, the most promising is teixobactin. This compound binds to two lipid components of bacterial cell walls, which stops their construction in its tracks. Moreover, neither lipid evolves rapidly. The slow mutation makes it likely that resistance will take much longer to arise than in traditional antibiotics.

Like the soil, the ocean teems with life. Drug scientists focus on one particular kind of aquatic life—microbes. They represent an immense, mostly untapped source of potential new antibiotics.

Also, like their earthbound counterparts, marine microbes are extremely difficult to culture. That's due to the vast differences in temperature, salinity, and pressure (among others) between ocean and lab. However, University of California San Diego scientists are tapping into the power of genomics to work around the difficulties. By isolating and sequencing the DNA of marine microbes, researchers are identifying gene clusters that they predict will code for antibiotic-like compounds. They then transfer the genes to bacteria that are easily grown in the lab. These bacteria then make the compound specified by the gene. One compound discovered by this technique, *taromycin*

A, impairs the growth of several types of drug-resistant bacteria.

On Beyond Antibiotics

In addition to creating new antibiotics, drug discovery scientists are taking entirely different tactics to combat bacterial infections in clinically relevant strains of bacteria such as drug-resistant *E. coli* and *Staphylococcus aureus*: bacteriophage treatments and CRISPR genome editing.

Pop Goes the Bacterium

Phage therapy turns one microbe against another. Drug researchers are coopting viruses that infect bacteria—bacteriophages—to kill them. The word "bacteriophage" comes from the Greek word *phagein*—"to devour." These voracious microbes often have a taste for the bacteria that harm us. Typically, a phage "partners" with only one strain.

The predator-prey drama unfolds like this: a phage attaches to a microbe's surface, punches holes in its membrane, and injects its own genetic material. The intruding phage replicates, creating so much new virus that the bacterium explodes. A slew of fresh viruses infects the other bacteria, eventually wiping them all out.

Bacteriophages and their infectious quarry have co-evolved for millennia, adapting and changing in response to one another. That's key. It means that humans are much less likely to develop resistance to phages—as we have too many antibiotics.

When researchers tweak a bacteriophage for therapeutic use, they select one that will attack only one specific type of *harmful* bacteria. This remarkable precision leaves many strains of benign bacteria that comprise our microbiome alone. Humans have lived safely with bacteriophage literally for ages, which suggests the viruses pose a minimal safety risk.

Because they're live viruses, the FDA considers bacteriophages a biologic treatment. The agency hasn't approved any phage treatments as of yet. The first clinical trial of a phage-based treatment for drug-resistant *Staphylococcus aureus* infections was initiated at University of California, San Diego in 2019.

Antivirals

As described in Chapter 1, viruses are very simple. They consist only of genetic material—DNA or RNA—surrounded by a protective protein coat and, in some cases, a lipid membrane. You might think that their unassuming construction makes them easy to target. In fact, the opposite is true. Their simple structure means that they rely on their victims—the host cells they've infected—to do what's required for them to survive. Like, make proteins, for example.

Viruses' absolute dependence on their hosts means that scientists must carefully study their lifecycle to figure out the precise points at which a therapy could interfere with the virus without harming the host cell. Typically, this means identifying a viral protein that plays a critical role and creating a small molecule drug to inhibit it.

The groundbreaking hepatitis C virus (HCV) drug, Sovaldi makes a great example of the development of a new antiviral. As a *nucleotide analog polymerase inhibitor*, Sovaldi blocks the viral polymerase—the enzyme HCV uses to replicate its genetic material (RNA). To copy the viral RNA, the polymerase simply connects new building blocks (nucleotides) in the same order as the existing viral RNA. The analog drug closely mimics the structure of naturally-occurring nucleotides, so the polymerase will subsequently incorporate it into a growing RNA strand. However, the analog drug has been chemically modified, so the polymerase can't add any additional nucleotides once it's been incorporated. This brings viral replication to a halt. Sovaldi is typically administered with Velpatasvir; a drug that inhibits the action of the viral protein NS5A. NS5A helps with viral replication. The combination is close to 100% effective. Prior to Sovaldi's approval, the only treatment available for HCV was interferon, which by itself was only effective in about 50% of cases.

CRISPR Genome Editing

CRISPR genome editing (described in detail in Chapter 6) may offer another route to highly specific, effective new anti-bacterial treatments. The idea here is to engineer bacteriophage to deliver CRISPR components, including guide RNA and the Cas3 protein, to their bacterial prey. The guide RNA directs the Cas3 protein to cut up the bacterial DNA.

CRISPR genome editing for people relies on the Cas9 protein. Cas9 cuts target DNA in a single, precise location. The related enzyme Cas3 chews up target DNA much more thoroughly than its cousin. This makes it better at destroying bacteria.

Genetic Disease

In reviewing treatments for genetic disease, we'll focus on **monogenic** disorders. These conditions are inherited through defects in a single gene. Geneticists can trace their pathology to flaws in the proteins the genes produce. Approaches to addressing these diseases vary:

- Replacement or correction therapies involve manufacturing a correct version of the faulty protein and delivering it as a biologic drug. These are often called enzyme replacement therapies (ERTs).
- Pharmaceutical chaperone therapies use a small molecule to help correct the protein.
- Direct genetic intervention through gene therapy or RNA-targeting therapies.

A **pharmacological chaperone** is a small molecule drug that mimics the action of a chaperone protein by serving as a scaffold for mutant proteins that would otherwise misfold. Pharmacological chaperones are designed to interact with a specific mutant protein, helping it achieve its correct shape.

Tricky Terms

Chaperone Protein
Most proteins fold into their final shape without assistance. Very large proteins or those with a highly complex structure need a little help to reach their final shape. Chaperone proteins serve as guides in this process. They protect large, complicated proteins from other proteins that might interact with them and hamper folding.

Replacement or Correction: Fabry Disease

Monogenic diseases affect any part or function of our bodies, from the immune system to red blood cell function to how cells handle waste products. For example, Fabry disease impairs the ability to process certain lipids (fats). Patients lack working versions of a critical enzyme, resulting in problems with the kidney, heart, and skin. The enzyme, galactosidase, helps break down glycolipids—fats with a carbohydrate attached. The Fabry mutation results in this enzyme being misfolded, which makes it useless.

Swapping Enzymes

In 2003, the FDA approved Fabrazyme, an enzyme replacement therapy. The supplemental galactosidase comes from Chinese hamster ovary (CHO) cells genetically engineered to make the protein, following standard biomanufacturing practices. The treatment is administered via infusion.

Pharmacological Chaperone: Migalastat

The small molecule drug Galafold also treats Fabry's disease. It, too, acts as a pharmacological chaperone, helping mutated galactosidase fold correctly. Galafold binds to galactosidase *during* the folding process, where it nudges the protein into the proper conformation. The now properly folded protein travels to the lysosomes, where it digests the lipids. The insides of these cellular compartments are acidic, which causes Galafold and the enzyme to come apart. The separation leaves behind functional galactosidase for the body to pick up and use.

To complicate matters, multiple mutations within the galactosidase gene cause Fabry disease. Galfold doesn't work on all of them. Galafold's manufacturer estimates that the drug will help 35% to 50% of patients.

Other pharmacological chaperones on the market include the cystic fibrosis drugs lumacaftor, tezacaftor, and elexacaftor. They act as chaperones to the protein CFTR, which causes cystic fibrosis. Each drug targets a different mutated version of the CFTR protein.

Genetic Therapies: Spinal Muscular Atrophy

Some monogenic conditions affect the nervous system. Spinal muscular atrophy (SMA) is a devastating genetic disease that destroys motor neurons. These neurons enable movement. In SMA's most severe form, babies often don't appear to move at all. Toddlers have trouble swallowing and breathing. The disease is usually fatal by age two.

SMA arises from mutations in the SMN1 gene, which produces the survival motor neuron (SMN) protein. Without it, the neurons work incorrectly and ultimately die. How quickly depends on the extent of the SMN protein deficiency. The less someone produces, the more severe their disease.

A closely related gene, the survival motor neuron 2 (SMN2) gene, produces a far smaller quantity of the protein. It serves as a sort of neurological backup. The amount of working SMN protein it produces preserves some motor nerve function. However, differences in the way SMN2 works mean most of its protein is non-functional and degrades very quickly.

Targeting the genetics of the disease has enabled researchers to come up with two powerful treatments that tackle the disease's root cause.

Exon Inclusion: The More Information, the Better

The first involves exon inclusion. If you recall, exons are sections of pre-mRNA that make it into mRNA, which contains the information a cell uses to produce its corresponding protein. The FDA approved the antisense drug Spinraza for SMA. Spinraza is a short, synthetic piece of RNA with a sequence directing it to bind to the SMN2 pre-mRNA. The Spinraza binding changes how cells process the SMN2 pre-mRNA. Specifically, it allows more information in the pre-mRNA to make it into the mRNA and get converted into protein. This results in more full-length, functional SMN protein to drive the motor neurons.

Gene Therapy: Special Delivery

In 2019, Zolgensma became the first commercially available gene therapy for SMA. Zolgensma uses a viral vector to deliver a corrected copy of the mutated SMN1 gene to affected babies. As a single gene disorder, SMA makes an ideal candidate for gene therapy. That's because delivering a "good" copy of the mutated gene to supply a permanent copy of the correct SMN1 protein-making instructions should cure the disease. Researchers used the AAV9 vector because, unlike most viral vectors, it can cross the blood-brain barrier.

Cancer

Although there are many different types of cancer, from leukemia to brain tumors to breast cancer, each arises from overactive cell division. In essence, our cells run away with themselves. They keep growing and dividing when they shouldn't, eventually crowding out healthy cells and tissue.

The treatments discussed below share the common feature of eliminating the extra cells—preventing them from dividing, killing them directly, or activating the immune system to attack them. Let's see how some of them work.

Small Molecule Disrupters

Small molecule oncology drugs work in two main ways: 1) shutting down growth factor signaling or 2) inducing apoptosis (cell suicide).

In *The Biotech Primer One* book the concept of growth factor signaling. It also discussed how the proteins involved in transmitting a growth signal from a cell-surface growth factor receptor to the cell nucleus make good targets for many cancer treatments. These drugs enter cells and interact directly with internal targets. Consequently, they can manipulate and inhibit the intracellular portion of growth factor receptors. They may also interfere with kinase enzymes that help transmit growth signals from the cell surface. Iressa and Tarceva are small molecule inhibitors of the epidermal growth factor receptor (EGFR). Both drugs treat non-small cell lung cancer.

Beneficial Suicide

Apoptosis, also known as programmed cell death or cell suicide, is a mechanism by which cells protect themselves from damage. Excessive DNA mutation or the buildup of damaged proteins can initiate apoptosis, which helps ensure that damaged cells won't become malignant. Many cancers are associated with defects in the process. For these, triggering apoptosis can effectively treat the disease.

Oncologists induce apoptosis in different ways. The first involves proteasomes. These cellular compartments, or organelles, dispose of unneeded or damaged proteins. Preventing proteasomes from cleaning up means damaged proteins accumulate. The pileup signals the cell that something is seriously amiss, setting off cell death—a welcome outcome in cancer. Small molecule drugs that activate apoptosis by reigning in proteasomes are Velcade and Krypolis. Both treat multiple myeloma.

Crippling cells' ability to repair DNA provides another way to trigger apoptosis. This is especially effective in cancers that stem from a defect in DNA repair enzymes. Faulty repair enzymes leaves those cells especially vulnerable to new mutations. BRCA1- and BRCA2-positive breast or ovarian cancers make an excellent example. When the BRCA proteins can't function, cells sustain damage to the genes that regulate their growth. Ultimately, this can lead to cancer.

Fortunately, breast and ovarian cells also rely on a second DNA repair protein, Poly (ADP-ribose) polymerase

(PARP). This enzyme provides an opening for healing. Researchers have learned that inhibiting PARP can trigger apoptosis when sufficient DNA damage occurs. Several small molecule PARP inhibitors are available, including Lynparza for breast and ovarian cancers, Rubraca and Zejula for ovarian cancer, and Talzenna for breast cancer. PARP inhibitors are typically administered with radiation or chemotherapy to increase cancer cell DNA damage and hence the PARP inhibitor's effectiveness. More DNA damage means apoptosis is more likely to happen.

Biologics: Monoclonal Antibodies and Company

When it comes to cancer biologics, monoclonal antibodies (mAbs) are the name of the game. As described in Chapter 4, mAbs work by an array of mechanisms. These include stopping growth signal transduction and revving up the patient's immune system.

Vectibix (for colorectal cancer), Erbitux (also for colorectal cancer), and Portrazza (for non-small cell lung cancer) block growth signals to tumor cells by binding to the growth factor receptor EGFR. They do so in a manner that doesn't activate the receptor but prevents the growth factor from binding to it and activating growth signaling.

Rituxan (for non-Hodgkin's lymphoma and chronic lymphocytic leukemia) treats blood cancers by recognizing and binding to the protein CD20 on the surface of B-cells (a type of white blood cell.) This process then activates cytotoxic T-cells, macrophages, and other white blood cells to destroy the cancerous B-cells.

Most oncological mAbs target proteins on the surface of a cancer cell—tumor antigens. This means that they're typically highly effective for one specific cancer, but not multiple cancers. The drug Avastin proves the exception. Instead of targeting one tumor antigen, it homes in on a mechanism at work in most tumors: angiogenesis. This is the growth of blood vessels into a tumor induced by the tumor's secretion of vascular endothelial growth factor (VEGF). The new blood vessels supply the oxygen and other nutrients the tumor needs. Without angiogenesis, tumor cells die. Avastin targets VEGF, binding, and sequestering it before it reaches its receptor on the surface of blood vessels. This disrupts their further growth into the tumor. Avastin is approved to treat a number of cancers, including colorectal, lung, breast, renal, brain, and ovarian.

Other immunotherapies—immune checkpoint inhibitors and CAR-T—directly activate cytotoxic T-cells to destroy cancer cells. Chapter 4 details these therapies' mechanisms of action. Immune checkpoint inhibitors on the market include Keytruda, Opdivo, Tecentriq, Imfinzi, and Bavencio. The FDA has approved two CAR-T therapies, Yescarta and Kymriah. Several more are in development.

Autoimmune Disorders

The precise causes of most autoimmune disorders remain unclear. Nonetheless, biotech companies have developed effective treatments by focusing on the global feature they share: an overactive immune system.

Biologics

The most common class of biologic drug for autoimmune disorders are TNF-alpha inhibitors. TNF-alpha is a strong pro-inflammatory cytokine—in other words, it promotes inflammation and white blood cell activation. This exacerbates the symptoms of autoimmune disorders. TNK-alpha is sometimes referred to as the "master regulator" of the inflammatory response. Activating or deactivating it also drives the response of other inflammatory cytokines. Many autoimmune patients overproduce TNF-alpha. mAb therapies designed to recognize, bind, and sequester any excess have proven to be invaluable at helping to control symptoms.

mAb Autoimmune Therapies

Several mAbs are on the market for autoimmune disorders, including Remicade (approved for Crohn's disease, ulcerative colitis, rheumatoid arthritis, psoriasis, psoriatic arthritis, and Behcet's disease), Humira (rheumatoid arthritis, psoriatic arthritis, ankylosing spondylitis, Crohn's disease, ulcerative colitis, psoriasis, hidradenitis suppurativa, uveitis, and juvenile idiopathic arthritis), Simponi (ulcerative colitis, rheumatoid arthritis, psoriatic arthritis, and ankylosing spondylitis), and Cimzia (Crohn's disease, rheumatoid arthritis, psoriatic arthritis, and ankylosing spondylitis.)

A related drug, Enbrel, also degrades excess TNF alpha. Instead of an antibody that recognizes TNF-alpha, it's a soluble version of the cytokine receptor. The therapeutic effect remains the same: mopping up surplus TNF-alpha. Enbrel treats rheumatoid arthritis, juvenile idiopathic arthritis, psoriatic arthritis, plaque psoriasis, and ankylosing spondylitis.

Small Molecules

As the previous paragraphs suggest, doctors, treat most serious autoimmune disorders with biologic drugs. However, some small molecule approaches exist.

Xeljanz treats rheumatoid arthritis and ulcerative colitis. It works by shutting down a kinase enzyme required for a cell-signaling pathway involved in activating white blood cells and releasing inflammatory cytokines. The FDA has also approved another small molecule drug, Tecfidera, for multiple sclerosis. Its exact mechanism is unknown, but it's also thought to modulate signaling pathways involved in activating white blood cells and releasing inflammatory cytokines.

Neurological Disease

As described in Chapter 1, neurological disorders cover a wide range of diseases, many of which aren't well understood or for which treatment options aren't available. Here we'll look at some novel therapies in development for Alzheimer's disease (AD), Huntington's disease (HD), and Parkinson's disease (PD).

Small Molecules

Preclinical studies suggest that the small molecule drug NNI-362 promotes the growth of new hippocampal neurons in mice. The hippocampus is one of the first regions of the brain to suffer damage in AD. It's what's thought to lead to the short-term memory loss and disorientation common in the early stages of disease. NNI-362 will shortly enter clinical trials, so researchers can determine if the effect carries through to humans.

Biologics

The precise mechanism of AD remains unknown. However, we do know the illness is associated with the build-up of amyloid-beta protein plaques in the brain. For many years, several companies pursued mAb therapies directed at amyloid-beta to remove these plaques. Unfortunately, none have so far been successful at improving the symptoms of AD. Antibodies targeting other proteins offer an alternative. For example, the mAb ANX005, now in preclinical development, seeks to sequester a protein known as c1q. This protein is thought to drive inflammation in the brain. Inflammation is considered to contribute to the development of AD.

RNA-Targeting Drugs

Currently, in Phase II clinical development for HD, RG6042 is an antisense drug designed to reduce the amount of mutated huntingtin protein produced in the brains of people with HD. Since the accumulation of mutated huntingtin protein is associated with disease progression, reducing the amount may well improve disease symptoms.

Gene Therapy

Researchers are also exploring gene therapy approaches to HD. One such product, AMT130, delivers a gene that codes for a microRNA targeting mRNA that provides the recipe for the huntingtin protein. microRNA dugs are very similar to the siRNA drugs described in Chapter 7. Their net effect is to reduce the amount of the targeted mRNA, which diminishes the amount of protein that mRNA codes for. Gene therapy's key advantage here is that it may provide "one and done" treatment, rather than requiring the repeated dosing of an antisense drug. AMT130 will soon enter clinical trials.

Why use antisense or microRNA therapy to diminish the amount of mutated protein in HD, rather than deliver a copy of the "good" gene sequence? The answer lies in the nature of the affliction. Current research suggests that HD arises from too much "bad stuff"—i.e., the mutated Huntingtin protein—rather than not enough "good stuff"—i.e., the correct huntingtin protein. Diseases such as hemophilia, for example, which result from the lack of a specific protein required for blood clotting, may be successfully treated by delivering "good" copies of a missing or dysfunctional protein. However, providing a corrected protein doesn't help when the protein itself causes the problem. Treating some diseases requires reducing or eliminating a defective protein altogether.

Gene therapy is also in clinical development for PD. VY-AADC01 delivers the AADC gene, which codes for an enzyme, L-amino decarboxylase (AADC). This enzyme is involved in dopamine production. PD patients experience lower levels of both AADC and dopamine. The makers of VY-AADC01 hope to restore dopamine to help relieve symptoms.

This chapter has looked at different types of diseases and potential therapeutic approaches. Although by no means comprehensive, we hope it provides an appreciation for the increasing number of therapeutic options available and some of the situations in which they apply.

Biopharma innovation is moving faster now than ever before. Sequencing individual patient's entire genomes, engineering white blood cells to attack specific cancers, and delivering corrective genes to treat genetic disorders all exemplify a technology that has evolved from long-shot to mainstream within a decade. These advances have been made possible by biotech researchers' unrelenting drive to understand the human body. It will continue to lead to more advances in human health and wellness. Thank you for inviting us to be part of your journey of discovery.

GLOSSARY

A

Adeno-Associated Virual Vector (AAV): A method by which genes can be inserted, modified, or deleted in organisms using adeno-associated virus. There are different serotypes of AAVs used as viral vectors for gene therapy.

Adjuvants: A substance or a combination of substances used to increase the efficacy or potency of certain drugs.

Amino Acid: Molecules that combine to form proteins. There are twenty different amino acids. The sequence of amino acids in a protein determines the protein's structure and function.

Angiogenesis: Growth of a network of blood vessels penetrates cancerous growths, supplying nutrients and oxygen and removing waste products.

Anion Exchange Chromatography: A type of column chromatography in which positively-charged proteins are captured by a negatively-charged chromatography matrix.

Antibody: A protein produced by the immune system that binds to a specific antigen.

Antibody-Drug Conjugates (ADCs): Highly targeted biologics that combine monoclonal antibodies, specific to surface antigens present on particular tumor cells, with highly potent anti-cancer agents linked via a chemical linker.

Antigen: A foreign substance which, when introduced into the body, stimulates an immune response.

Antigen Presenting Cell: A highly specialized cell that is able to process antigens and display their peptide fragments on the cell surface together with molecules required for lymphocyte activation.

Antisense: The non-coding DNA strand of a gene.

Antiserum: The fluid component of clotted blood from an immune individual that contains antibodies against the molecule used for immunization.

Apoptosis: The process of cell self-destruction.

Aseptic: Free from contamination caused by harmful bacteria, viruses, or other microorganisms.

Assay: A test.

Atom: A particle, made up of a nucleus and one or more orbiting electrons, which is the basic unit of a chemical element.

Adenosine Triphosphate (ADP): A compound used by cells to store energy and to fuel metabolic processes.

Autoimmune Disease: A condition in which your immune system mistakenly attacks your body.

B

B-Cell: An antibody-producing cell of the immune system.

B-Cell Receptor: The cell-surface receptor of B- cells for a specific antigen.

Bacterium: A group of small, single-celled, prokaryotic microorganisms.

Base: One of the molecules - adenine, guanine, cytosine, thymine, or uracil - which form part of the structure of DNA and RNA molecules. The order of bases in a DNA molecule determines the structure of proteins encoded by that DNA. See nucleotide.

Base Pair (bp): Two complementary nucleotide bases joined together by chemical bonds. The base adenine pairs with thymine, and guanine pairs with cytosine.

Basophil: White blood cells that participate in the inflammatory response.

Bioactive: Having an effect on a biological system.

Bioavailability: The proportion of a drug or other substance which enters the circulation when introduced into the body and so is able to have an active effect.

Biologic: Products of living organisms or cells used in the treatment or management of a disease.

Biologics Licensing Application (BLA): An application for marketing approval for a biologic drug; submitted to the FDA upon successful completion of Phase III clinical trials.

Biomanufacturing: The use of living cells to produce a biological product. An example is a therapeutic protein.

Biomarker: A physiological event or molecule that can be measured. Examples include the presence or absence a protein or a mutated gene. Biomarkers are often used to indicate the presence or progression of a disease.

Biopharmaceutical: Drugs, either chemical compounds or biologics, made using the method of rational drug design.

Bioreactor: The usually stainless steel tank used to grow living cells that produce biologics. A bioreactor can range in size from a few liters up to 100,000 liters.

Biosciences: A term meant to encompass both biotech and pharmaceutical companies.

Biosimilar: A biologic deemed highly similar to an already approved biologic medicine, known as a reference product. This similarity is confirmed through a number of tests, both in the lab and in clinical research.

Biotechnology: The use of cellular and biomolecular processes to solve problems and make useful products

C

Carriers: A person who has inherited a genetic trait or mutation but who does not display that trait or show symptoms of the disease.

C-C Chemokine Receptor Type 5 Gene (CCR5Gene): A gene that codes for a protein on the surface of white blood cells that is involved in the immune system as it acts as a receptor for chemokines. Many forms of HIV, the virus that causes AIDS, initially use CCR5 to enter and infect host cells.

Cell: The basic subunit of any living organism, typically containing at a minimum genetic material, an energy-producing system, and protein-making machinery, all surrounded by a membrane.

Cell Bank: A uniform population of cells, stored under defined conditions, typically frozen at -80 degrees Celsius or colder. The assumption is that each vial of cells is comparable, and may be used in a consistent manner after being thawed. See Master Cell Bank and Working Cell Bank.

Cell Line: A cell culture developed from a single cell and therefore consisting of cells with a uniform genetic makeup.

Cell Lysate: The cellular debris and fluid produced by breaking open (lysing) a cell.

Cell Membrane: Barrier made out of fats and proteins that separates the inside of the cell from the outside.

Cell Wall: A stiff covering around the cell membrane of certain non-animal cells, such as plants and many types of bacteria.

Channel Protein: A protein that spans the cell membrane, allowing substances to pass through form the outside to the inside of the cell.

Chaperone Proteins: Proteins that assist with the conformational folding or unfolding and the assembly or disassembly of other proteins.

Checkpoint Inhibitor: Are key regulators of the immune system that when stimulated can dampen the immune response to an immunologic stimulus. Some cancers can protect themselves from attack by stimulating immune checkpoint targets.

Chemokine: Signaling molecules that are involved in the activation and migration of immune system cells. Chemokine signaling plays a key role in the inflammatory response.

Chimeric Antibody: Antibodies whose gene sequence consists of DNA from two different species. Typically, the term refers to antibodies whose DNA is between 10% and 25% mouse origin, with the remaining sequence being of human origin.

Chimeric Antigen Receptor T-Cells (CAR-T): Are T-cells that have been genetically engineered to fight cancer by changing them in the lab so they can find and destroy cancer cells

Chromosome: A long strand of DNA found within cells. Chromosomes contain both genes and regions of DNA that do not code for proteins.

Chromosomal Abnormalities: Result from mutations that change the number of chromosomes (numerical abnormalities) or change the chromosome structure (structural abnormalities). They may alter the ability of the cell to survive and function.

Cleave: To split a molecule by breaking a particular chemical bond.

Clinical Endpoint: In a clinical research trial, a clinical endpoint generally refers to occurrence of a disease, symptom, sign or laboratory abnormality that constitutes one of the target outcomes of the trial.

Clonal Selection: A process proposed to explain how a single B or T cell that recognizes an antigen that enters the body is selected from the pre-existing cell pool of differing antigen specificities and then reproduced to generate a clonal cell population that eliminates the antigen.

Codon: A set of three nucleotide bases in a DNA or RNA sequence, which together code for a unique amino acid. For example, AUG (adenine, uracil, guanine) codes for the amino acid methionine.

Combination Product: A product comprised of two or more FDA-regulated categories.

Combinatorial Chemistry: A technique for rapidly and systematically assembling different combinations of molecules to create tens of thousands of diverse compounds. Used in drug discovery screening assays to identify potential useful therapeutic candidates.

Combinatorial Libraries: Collections of chemical compounds, small molecules or macromolecules such as proteins, synthesized by combinatorial chemistry, in which multiple different combinations of related chemical species are reacted together in similar chemical reactions.

Companion Diagnostic: A diagnostic used by a physician to inform his prescribing decision.

Contract Research Organization (CRO): A company that conducts preclinical or clinical trials for another company on a contract basis.

Control Group: The group in an experiment or study that does not receive the experimental treatment and is then used as a benchmark to measure how the experimental group does.

Current Good Manufacturing Practices (cGMP): FDA guidelines governing the manufacturing of biopharmaceuticals.

Cytochrome P450: A liver enzyme which plays an important role in metabolizing drugs.

Cytokine: Proteins made by cells that affect the behavior of other cells.

Cytosine: One of the four nucleotide bases that make up DNA.

Cytotoxic T-Cells: T-cells that can kill other cells, typically virus-infected cells or tumor cells.

Cytotoxins: Proteins made by cytotoxic T-cells that participate in the destruction of other cells.

D

Deletion Mutation: One or more nucleotides removed from a DNA sequence during the replication process.

Denature: Dramatic change in the conformation of a protein, usually unfolding caused by heat or exposure to chemical such as acids. Most often will result in loss of biological function.

Deoxyribonucleic Acid (DNA): The molecule that encodes genetic information. DNA is a double-stranded helix held together by bonds between pairs of nucleotides.

Diagnostic (Dx): A test used to identify a disease or disorder, or to monitor the progression of treatment. Routine diagnostics are broad screening tools, whereas a specialty diagnostic screens for a specific disease.

Directed Evolution: A method used in protein engineering that mimics the process of natural selection to steer proteins or nucleic acids toward a user-defined goal.

DNA Sequence: The precise order of nucleotides within a DNA molecule. It includes any method or technology that is used to determine the order of the four bases—adenine, guanine, cytosine, and thymine—in a strand of DNA.

Dominant: A disease-associated gene is dominant if only one copy of it needs to be present to cause disease.

Double Helix: The shape of the DNA molecule.

DNA Polymerase: The enzyme that replicates DNA.

DNA Repair Genes: A collection of genes that code for proteins that enable a cell to identify and correct damage to the DNA molecules that encode its genome.

Domain: Portion of a protein that has its own structure.

Drug Candidate: A small molecule or biologic that is being tested for its therapeutic potential.

Drug Development: The process of testing therapeutic molecules for safety and efficacy in animals and humans, and developing appropriate formulation, delivery, and manufacturing methods.

Drug Discovery: The process of identifying molecules with a therapeutic effect against a target disease.

Drug Target: Organ, tissue, or molecule involved in a disease that is modified or affected by a potential therapeutic.

E

Efficacy: The ability of a substance to produce a desired clinical effect.

Epidermal Growth Factor Receptor (EGFR): Transmits proliferation signals to many different cell types. Mutations in this protein are associated with different types of cancer.

Electroporation: The act of applying an external electrical field to a cell membrane in order to increase its permeability. The technique is used as a way of introducing DNA into bacterial cells.

Enzyme: A protein that enables a biochemical reaction in a cell.

Eosinophil: White blood cells thought to be important chiefly in the defense against parasitic infections.

Epidermal Growth Factor (EGF): A protein which binds to the epidermal growth factor receptor and stimulates cell division.

Epigenetic: Relating to or arising from nongenetic influences on gene expression.

Epitope: The structural feature of an antigen molecule to which an antibody binds.

Erythrocyte: Red blood cells.

Eukaryote: Cell or organism with mem- brane-bound, structurally discrete nucleus and other well-developed subcellular compartments.

European Medicines Agency: European Union agency for the evaluation and supervision of medicinal products.

Evolution: The natural selection of beneficial changes.

Excipient: Anything present in a drug product that is not the active ingredient. Includes components such as bulking and stabilizing agents, preservatives, salts, solvents, and water.

Ex vivo: Pertaining to a biological process or reaction taking place outside of a living organism.

Exon: Coding sections of an RNA transcript, or the DNA encoding it, that are translated into protein.

F

Formulary: An official list giving details of medicines that may be prescribed.

Formulation: The process in which different chemical substances, including the active drug, are combined to produce a final medicinal product.

G

G Protein-Coupled Receptor (GPCR): A type of receptor protein that is targeted in over 25% of all biotech drugs.

Gamete: Reproductive cells; the ovum or sperm.

Gene: A length of DNA that codes for a particular protein.

Gene Amplification: The multiple replication of a section of the genome, which occurs during a single cell cycle and results in the production of many copies of a specific sequence of the DNA molecule.

Gene Expression: The process by which the information in a gene is used to create proteins.

Gene Of Interest: A gene that is under study by a researcher as a potential recipe for a biologic drug or because of its relevance to disease.

Gene Product: The protein produced by a gene.

Gene Therapy: A medical field which focuses on the genetic modification of cells to produce a therapeutic effect or the treatment of disease by repairing or reconstructing defective genetic material.

Generally Regarded As Safe (GRAS): A special status afforded by the FDA to ingredients and methods that have a proven, longstanding history of causing no harm to humans or animals.

Generic Drug: A small molecule drug that is an exact molecular copy of an innovator drug.

Genetic Code: Set of rules by which the information encoded in DNA is translated into proteins.

Genetic Engineering: Altering the genetic material of cells or organisms in order to make them capable of producing new substances or performing new functions.

Genetic Variation: Differences in DNA sequence that occurs between individuals.

Genetically Engineered Organism: An organism whose DNA has been altered using genetic engineering techniques.

Genome: All of the genetic material in the chromosomes of a particular organism.

Genome Editing: Manipulation of the genetic material of a living organism by deleting, replacing, or inserting a DNA sequence, typically to correct a genetic disorder.

Genomics: The branch of molecular biology concerned with the structure, function, evolution, and mapping of genomes.

Genotype: The genetic constitution of an individual organism.

Germ Cell: A cell containing half the number of chromosomes of a somatic cell and able to unite with one from the opposite sex to form a new individual; a gamete.

Glycosylation: Adding one or more carbohydrate molecules onto a protein after it has been built by the ribosome.

Golgi Body: Cellular organelle which sorts and sends proteins to their appropriate location within the cell.

Granulocyte: A type of white blood cells characterized by the presence of small particles in their cytoplasm.

Growth Factor: A substance, such as a vitamin or hormone, which is required for the stimulation of growth in living cells.

Growth Factor Signaling: The process of transmitting a chemical signal from a cell surface receptor to the nucleus of a cell.

Growth Medium: A liquid or gel designed to support the growth of cells or microorganisms. There are different types of media for growing different types of cells.

Guanine: One of the four nucleotide bases that makes up DNA.

H

Half-Life: The amount of time it takes for 50% of a drug given to a patient to be eliminated or destroyed by natural processes.

Haplotype: Set of single nucleotide polymorphisms (SNPs) on a chromosome which are statistically associated.

Hematopoiesis: The generation of the cellular components of blood, including red blood cells, white blood cells, and platelets.

Hematopoietic Stem Cell (HSC): Stem cells found in the bone marrow that have the potential to develop into any of the different types of blood cells found in the body.

Hemostasis: Any self-regulating process by which biological systems tend to maintain stability while adjusting to conditions that are optimal for survival.

Histone: Proteins that chromosomal DNA is wrapped around.

Host: An animal or plant on or in which a parasite or commensal organism lives.

Human Anti-Mouse Antibody (HAMA): Antibodies generated by the human immune system when a human is injected with an antibody produced by a mouse.

Humanized Antibody: An antibody produced in a non-human species whose DNA sequence has been altered to make it more closely resemble a human antibody.

Hybridoma: Hybrid cell lines formed by fusing a specific antibody-producing B cell with a myeloma cell line.

i

Immune System Checkpoints: Regulators of the immune system. These pathways are crucial for self-tolerance, which prevents the immune system from attacking cells indiscriminately. Inhibitory checkpoint molecules are targets for cancer immunotherapy due to their potential for use in multiple types of cancers.

Immunogenicity: The ability of a foreign substance, such as an antigen, to provoke an immune response in the body of a human or other animal.

Immunological Memory: The ability of the immune system to quickly and specifically recognize an antigen that the body has previously encountered and initiate a corresponding immune response.

Infectious Disease: Sickness caused by organisms such as bacteria, viruses, fungi, or parasites.

Inflammation: A non-specific immune defense by the body in response to injury or the presence of foreign particles.

Innate Immune Response: The early phase of the host response to infection in which a variety of mechanisms recognize and respond to a pathogen. Innate immunity is present in all individuals at all times, does not increase with exposure to a given pathogen, and does not discriminate between pathogens. Inflammation is an example of an innate immune response.

Innate Immune System: The "first line of defense" in immunity. Responds non-specifically to any invading pathogen.

Innovator Drug: The product first authorized worldwide for marketing (as a patented product) based on its efficacy, safety, and quality, according to requirements at the time of authorization

Insertional Mutagenesis: The creation of DNA mutations by adding one or more base pairs. This can occur naturally or can be artificially for research purposes in the lab.

Interferon: Cytokines that can induce cells to resist viral infection.

Interleukin: A generic term for cytokines produced by white blood cells. Specific interleukins may have either an inhibitory or a stimulatory effect on the immune system.

Intron: Non-coding regions of an RNA transcript, or the DNA encoding it, that are eliminated by splicing before translation.

Ions: A positively or negatively charged atom.

Ion Channel: A channel protein through which ions are allowed to pass.

In Silico: Studies done on a computer; for example, modeling the structure and function of a protein.

In Vitro: Pertaining to a biochemical process or reaction taking place in a test-tube as opposed to taking place in an organism.

In Vivo: Pertaining to a biological process or reaction taking place in a living organism.

Insertion Mutation: One or more nucleotides is added to the DNA sequence.

Investigational New Drug (IND): A drug which has gained FDA approval to be shipped across state lines, typically for clinical trials, but has not yet gained approval for marketing.

K

Kinase: An enzyme that transfers a phosphate group from ATP to a protein. Often, this results in activation of the recipient protein.

Knockin Mice: A mouse whose DNA is genetically engineered by adding a gene sequence that is not typically found within the mouse so that it can express a particular protein.

Knockout Mice: A mouse whose DNA is genetically engineered so that it does not express particular proteins.

L

Large Molecule Drug: Another name for protein therapeutic or biologic. Large molecule drugs are too large to enter cells.

Lentiviral Vector: A method by which genes can be inserted, modified, or deleted in organisms using lentivirus.

Leukocyte: General term for white blood cell.

Life Sciences: A term meant to encompass both biotech and pharmaceutical companies.

Ligand: Any molecule that binds to a specific site on a protein or other molecule.

Ligation: The act of "gluing" two pieces of DNA, using the enzyme DNA ligase.

Liposome: Artificial lipid bilayer vesicle.

Lymphocyte: A white blood cell that is important in the body's immune response.

Lymph System: The system of lymphoid channels that drains extracellular fluid from the periphery.

M

Macrophage: A type of white blood cells that destroys pathogens by engulfing them.

Master Cell Bank (MCB): A culture of fully characterized cells distributed into separate vials, processed together in such a manner as to ensure uniformity. The master cell bank is usually stored at -80 degrees or colder (liquid nitrogen), and at two geographically distinct locations.

Maximum Tolerated Dose (MTD): Highest dose of a pharmacological treatment that will produce the desired effect without unacceptable toxicity.

Mechanism-Based Drug Design: The development of new therapeutics based on an understanding of the underlying disease mechanism.

Medical Device: An instrument, apparatus, implant, in vitro reagent, or other similar or related article, which is intended for use in the diagnosis of disease or other conditions, or in the cure, mitigation, treatment, or prevention of disease, or intended to affect the structure or any function of the body and which does not achieve any of its primary intended purposes through chemical action within or on the body.

Memory B-Cell: A B-cell that remembers the same pathogen for faster antibody production in future infections.

Messenger RNA (mRNA): The DNA of a gene is copied into mRNA molecules, which then serve as a template for the synthesis of proteins.

Microbiome: The totality of microbes, their genomes, and environmental interactions in a particular environment.

Microinjection: The use of a glass micropipette to inject a liquid substance at a microscopic or borderline macroscopic level. The target is often a living cell but may also include intercellular space.

Mitochondrion: An organelle within a cell that generates most of the cell's energy.

Molecular Cloning: A set of experimental methods in molecular biology that are used to assemble recombinant DNA molecules and to direct their replication within host organisms. The use of the word cloning refers to the fact that the method involves the replication of one molecule to produce a population of cells with identical DNA molecules.

Molecule: Two or more atoms connected by chemical bonds.

Monoclonal Antibody (mAb): An antibody produced by a single clone of cells, which therefore consistently binds to the same epitope of an antigen.

Monocyte: Precursor to macrophages.

Monogenic Disease: A disease that can be linked to a mutation in one specific gene.

Multicellular: Having or consisting of many cells.

Multiple Ascending Doses (MAD): A group of patients receives multiple low doses of the drug, while samples (of blood, and other fluids) are collected at various time points and analyzed to acquire information on how the drug is processed within the body. The dose is subsequently escalated for further groups.

Mutation: A change, deletion, or rearrangement in the DNA sequence that may lead to the synthesis of an altered or inactive protein or the loss of the ability to produce the protein.

Myeloma: A cancer of plasma cells.

N

Natural Selection: A mechanism of evolution whereby members of a population with the most successful adaptations to their environment are most likely to survive and reproduce.

Neoantigens: Newly formed antigens that have not been previously recognized by the immune system.

Neutrophil: Help fight infection by ingesting microorganisms and releasing enzymes that kill the microorganisms. A type of white blood cell, a type of granulocyte, and a type of phagocyte.

Neurotransmitters: A type of chemical messenger which transmits signals across a chemical synapse, such as a neuromuscular junction, from one neuron (nerve cell) to another "target" neuron, muscle cell, or gland cell.

New Drug Application (NDA): An application for marketing approval for a small molecule drug; submitted to the FDA upon successful completion of Phase III clinical trials.

Non-Specific Immune Response: A non-specific immune cell is an immune cell (such as a macrophage, neutrophil, or dendritic cell) that responds to any foreign threat. Non-specific immune cells function in the first line of defense against infection or injury.

Nuclear Membrane: The membrane surrounding the nucleus.

Nucleic Acid: One of the family of molecules which includes the DNA and RNA molecules.

Nucleotide: The "building block" of nucleic acids, such as DNA and RNA molecules. A nucleotide consists of one of five bases - adenine, guanine, cytosine, thymine, or uracil - attached to a sugar-phosphate group.

Nucleus: The membrane bound structure containing a cell's DNA found within all eukaryotic cells.

Nerve Growth Factor: A growth factor that binds to receptors on the surface of nerve cells, stimulating their growth.

Neutrophil: A major class of white blood cells. Play an important role in engulfing and killing foreign invaders.

O

Oncogene: A gene which in certain circumstances can transform a cell into a tumor cell.

Oncolytic Virus: A virus that has been genetically engineered to selectively infect and kill cancer cells.

Organelle: Membrane-bound structures in a cell that have specialized functions, such as mitochondria and the nucleus.

Orphan Drug: A drug developed for a condition that affects fewer than 200,000 individuals in the US.

Osteoblast: A cell responsible for bone formation.

P

p53: A tumor suppressor gene. Mutations in p53 are linked to many different types of cancer.

Pathogen: A bacterium, virus, or other micro-organism that can cause disease.

Peptide Bond: A type of chemical bond connecting amino acids.

Peptide Vaccine: A peptide-derived semi-synthetic vaccine.

pH: A measure of the acidity (hydrogen ion concentration) of a solution.

Phagocyte: A type of cell within the body capable of engulfing and absorbing bacteria and other small cells and particles.

Pharmacodynamics (PD): The study of the effect of a drug on the body; in particular, the effect of the drug as it relates to increasing dose.

Pharmacogenomics: The study of the role of genetics in drug response. It deals with the influence of acquired and inherited genetic variation on drug response in patients by correlating gene expression or single-nucleotide polymorphisms with drug absorption, distribution, metabolism and elimination.

Pharmacokinetics (PK): The study of drug absorption, drug distribution within the body, drug metabolism, and drug excretion.

Pharmacopoeia: An official publication containing a list of medicinal drugs with their effects and directions for their use.

Pharmacovigilance: The pharmacological science relating to the collection, detection, assessment, monitoring, and prevention of adverse effects with pharmaceutical products.

Phase I Clinical Trial: Initial studies to determine the metabolism and pharmacologic actions of drugs in humans, the side effects associated with increasing doses, and to gain early evidence of effectiveness; may include healthy participants and/or patients.

Phase II Clinical Trial: Controlled clinical studies conducted to evaluate the effectiveness of the drug for a particular indication or indications in patients with the disease or condition under study and to determine the common short-term side effects and risks.

Phase III Clinical Trial: Expanded trials after preliminary evidence suggesting effectiveness of the drug has been obtained, and are intended to gather additional information to evaluate the overall benefit-risk relationship of the drug and provide and adequate basis for physician labeling.

Phase IV Clinical Trial: Clinical trials conducted to identify and evaluate the long-term effects of new drugs and treatments over a lengthy period for a greater number of patients. Phase IV research takes place after the FDA approves the marketing of a new drug. Through Phase IV clinical studies, new drugs can be tested continuously to uncover more information about efficacy, safety and side effects after being approved for marketing.

Phosphorylation: The process of introducing a phosphate group into a molecule or compound.

Plasma Cell: An antibody-producing B-cell.

Plasmid: A small, circular piece of DNA that is separate from the cell's genome. Plasmids are manipulated in the laboratory to deliver specific genetic sequences into a cell.

Platelets: Blood cells that are required for blood clotting.

Point Mutation: A single nucleotide change in a DNA sequence.

Polyclonal Antibody: A mixture of antibodies that recognize different epitopes on the same antigen; each antibody is produced by a different B-cell.

Polygenic Disease: A disease that results from interactions among two or more genes.

Polypeptide Chain: A chain of amino acids linked together through chemical bonds.

Prostaglandins: A group of lipids made at sites of tissue damage or infection involved in dealing with injury and illness. They control processes such as inflammation, blood flow, the formation of blood clots, and the induction of labor.

Post-Translational Modification: Changes made to a protein after it has been made by a cell.

Precision Medicine: A medical model that proposes the customization of healthcare, with medical decisions, practices, and/or products being tailored to the individual patient. In this model, diagnostic testing is often employed for selecting appropriate and optimal therapies based on the context of a patient's genetic content or other molecular or cellular analysis.

Preclinical Studies: The testing of experimental drugs in the test tube or in animals - the testing that occurs before trials in humans may be carried out.

Primary Response: The adaptive immune response after initial exposure to an antigen.

Product Pipeline: The series of products developed and sold by a company, ideally in different stages of their life cycle.

Progenitor Cells: The more differentiated offspring of stem cells that give rise to distinct subsets of mature blood cells.

Prokaryote: Cell or organism lacking a membrane-bound, structurally discrete nucleus and other subcellular compartments. Bacteria are prokaryotes.

Promoter: A segment of DNA located in front of a gene, which provides a site where an enzyme can bind to the DNA molecule, to initiate transcription.

Prostaglandins: Signaling molecules that have a variety of physiological effects, including smooth muscle contraction, platelet aggregation, and control of cell growth.

Protease: An enzyme that degrades proteins.

Protein: A biological molecule that consists of many amino acids linked together by peptide bonds. As the chain of amino acids is being synthesized, it is also folded into higher order structures. Proteins are required for the structure, function, and regulation of cells, tissues, and organs in the body.

Proto-Oncogene: Any gene capable of becoming a cancer-producing gene (an oncogene). Proto-oncogenes have important functions in the normal cell, but, by mutation or by the acquisition of genetic control elements from oncoviruses they can lose their normal regulatory functions and lead to uncontrolled multiplication.

PTEN: A human tumor suppressor proteins. As such, it regulates cell growth and division.

R

R Group: In amino acids, the chemical group that varies and gives each amino acid its unique chemical and physical properties.

Randomized Double Blind Studies: Patients are randomly assigned to either the treatment or control group, and neither the patient nor the trial administers know which group is receiving the treatment.

Rational Drug Discovery: The development of new therapeutics based on an understanding of the underlying disease mechanism.

Real-World Evidence (RWE): Evidence obtained from real-world data (RWD), which are observational data obtained outside the context of randomized controlled trials (RCTs) and generated during routine clinical practice.

Receptor: A protein usually found on the surface of a cell that binds to a specific chemical messenger, such as a neurotransmitter or hormone.

Recessive: A disease-associated gene is recessive if two abnormal copies need to be present to cause disease.

Recombinant DNA (rDNA): DNA molecules that have been created by combining DNA from more than one source.

Reference Product: The single biological product, already approved by FDA, against which a proposed biosimilar product is compared.

Regulatory T-Cells: A subpopulation of T cells that modulate the immune system, maintain tolerance to self-antigens, and prevent autoimmune disease.

Regulome: Refers to the whole set of regulatory components in a cell. Those components can be regulatory elements, genes, mRNAs, proteins, and metabolites.

Research Support Company: A company who does not actively engage in drug discovery, but supplies all of the tools and technologies required.

Restriction Enzyme (RE): Proteins that recognize and cut DNA strands like molecular scissors along specific gene sequences.

Resolution: The dissipation of an inflammatory response.

Ribonucleic Acid (RNA): A nucleic acid similar to DNA but based on the sugar ribose and containing the nucleotides guanine, adenine, uracil, and cytosine instead of guanine, adenine, thymine, and cytosine, and typically single-stranded.

Ribosome: The cell structures within which protein synthesis occurs.

RNA Interference (RNAi): A technique used to block the expression of a particular protein. Also referred to as RNA inhibition.

S

Scale-Up: The process of slowly increasing the volume of a cell culture from a few milliliters to several thousand liters.

Secondary Response: The antibody response induced by a second exposure to the antigen.

Sequence: The order of nucleotides in a DNA or RNA molecule, or the order of amino acids in a protein.

Signal Transduction: The general process by which cells receive information from their environment.

Single Ascending Doses (SAD): Subjects are dosed in small groups called cohorts. Each member of a cohort might receive a single dose of the study drug or a placebo. A very low dose is used for the first cohort. The dose is then escalated in the next cohort if safety and tolerability allow. Dose escalation is stopped when maximum tolerability and/or maximum exposure is reached.

Single Nucleotide Polymorphism (SNP): A difference in one base pair between two DNA sequences.

Small Interfering RNA (siRNA): A class of double-stranded RNA molecules, 20-25 base pairs in length, which operate within the RNA inhibition pathway.

Small Molecule Drug: A drug that is chemically synthesized in the lab. Small molecule drugs are small enough to enter cells.

Somatic Cell: A cell that is not an egg or a sperm cell.

Specific Immune Response: The immune response triggered by a specific pathogen.

Statins: A class of drugs often prescribed by doctors to help lower cholesterol levels in the blood. By lowering the levels, they help prevent heart attacks and stroke. Studies show that, in certain people, statins reduce the risk of heart attack, stroke, and even death from heart disease by about 25% to 35%.

Start Codon: The three-nucleotide sequence that signifies the ribosome to start translating an mRNA sequence into a protein.

Stop Codon: The three-nucleotide sequence that signifies the ribosome to stop translating an mRNA sequence into a protein.

Stromal Cells: Connective tissue in the bone marrow and other tissue types.

Substitution Mutation: A single nucleotide is exchanged for a different nucleotide in a DNA sequence.

Substrate: A molecule on which enzymes act.

Susceptibility Gene: A genetic alteration that increases an individual's predisposition to a particular disease or disorder. When such a mutation is inherited, the development of symptoms is more likely, but not certain.

Surrogate Endpoint: In clinical trials, a surrogate endpoint (or surrogate marker) is a measure of effect of a specific treatment that may correlate with a real clinical endpoint but does not necessarily have a guaranteed relationship.

T

T-Cell: An immune system cell that recognizes specific pathogens based on the shape of its cell-surface receptor.

T-Cell Dependent Activation: Certain antigens can only activate B-cells with the help of T-cells.

T-Cell Receptor: Protein on the surface of T-cells to which antigen binds, activating T-cell.

T-Helper Cell: T-cells that help to fully activate antibody-secreting plasma cells by secreting activating cytokines.

Target Validation: Determining if targeting a particular molecule thought to be involved in a disease mechanism will be a safe and effective means of therapy.

Template: The strand of DNA that is being used by the DNA polymerase to construct a new DNA molecule.

Tetramer: A protein composed of four subunits.

Therapeutic Window: Concentration range within which a drug is both effective and safe.

Thrombocytes: Blood cells that are required for blood clotting. Also called platelets.

Thymine: A compound which is one of the four constituent bases of nucleic acids. It is paired with adenine in double-stranded DNA.

Thymus: A small, irregular-shaped gland in the top part of the chest, just under the breastbone and between the lungs. T-cells complete their development in the thymus.

Tissue Culture: The growth in an artificial medium of cells derived from living tissue.

Transcription: The process during which the information in a length of DNA is used to construct an mRNA molecule.

Transfection: The process of introducing DNA into eukaryotic cells.

Transfer RNA (tRNA): RNA molecules that bind to amino acids and transfer them to ribosomes, where protein synthesis is completed.

Transformation: A process by which the genetic material carried by an individual cell is altered by incorporation of external DNA into its genome.

Transgenic Organism: An organism whose genome has been altered by the incorporation of foreign DNA.

Translation: The process during which the information in mRNA molecules is used to construct proteins.

Tregitope: An epitope recognized by regulatory T-cells.

Tumor Microenvironment: The environment around a tumor, including the surrounding blood vessels, immune cells, fibroblasts, signaling molecules, and the extracellular matrix.

Tumor Suppressor Gene: A gene that protects a cell from one step on the path to cancer. When this gene mutates to cause a loss or reduction in its function, the cell can progress to cancer, usually in combination with other genetic changes.

U – Z

Unicellular: Organisms made up of only one cell.

Uracil: One of the four nucleotide building blocks of DNA.

Vasodilation: Dilation of a blood vessel.

Vector: A vehicle for the transfer of DNA from one organism to another. A plasmid is a common type of vector.

Viral Vector: Tools commonly used by molecular biologists to deliver genetic material into cells. The tool in this case is a virus.

Virus: An infective agent that typically consists of a nucleic acid molecule in a protein coat and can multiply only within the living cells of a host.

Virus-Like Particle (VLP): Molecules that closely resemble viruses but are non-infectious because they contain no viral genetic material.

Working Cell Bank (WCB): A cell bank that is established from one of the master cell bank vials.

X-Ray: A form of electromagnetic radiation.

Xenotransplantation: The therapeutic use of animal organs in people.

Made in the USA
Middletown, DE
24 September 2023